BBC

THE STORY OF THE

SOLAR SYSTEM

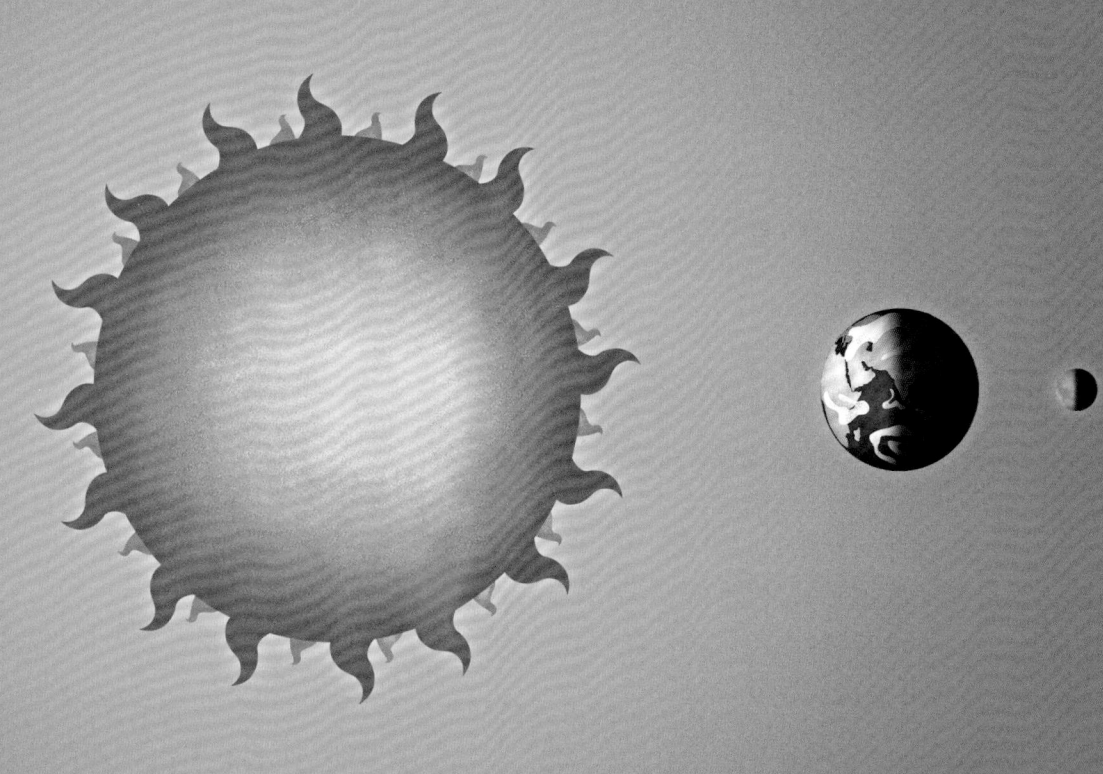

BBC

THE STORY OF THE
SOLAR SYSTEM

A VISUAL JOURNEY

DR MAGGIE ADERIN-POCOCK

WITH

SIMON GUERRIER

DESIGNER AND ILLUSTRATOR

EMMA PRICE

CONTENTS

WARNING: NOT TO SCALE

FOREWORD

SINCE WE, AS HUMANS, FIRST LOOKED UP AT THE NIGHT SKY AND WONDERED WHAT WAS OUT THERE, WE HAVE BEEN GATHERING INFORMATION ABOUT WHAT WE SEE.

It is one of the things that we share culturally across the globe: a desire to get a better understanding of the night sky. Astronomy, therefore, is often considered to be the oldest science. Early information gathered was passed on via stories, myths and legends. With time, these traditions evolved into more permanent forms of recording, in what we might now call early forms of data. The Lembombo bone, found in South Africa, is around 40,000 years old according to carbon dating, and its markings are thought to indicate a lunar calendar. The Nebra disk, found in Germany and dating from circa 1,600 BCE, shows an inverted image of the Sun, Moon and Milky Way on a bronze disk, while a Babylonian tablet from circa 164 BCE records the passing of what we now know as Halley's comet. These are just a few examples of early astronomical data recording. Today, we continue to gather information on things that are out there in many different forms.

Over the years, as we have gathered more data, our understanding of the cosmos has evolved. We started by looking just with our eyes and, seeing the Sun and Moon rise and set, we concluded that the Earth was the centre of the Universe. Yet irregularities were also noticed – objects that did not follow and flow with the others. These were called wandering stars, and they seemed to backtrack on themselves as they travelled across the sky. Further, more detailed and accurate observations revealed that these wandering stars were indeed planets, and they did not orbit around us but around the Sun.

There have been giant leaps in our understanding, such as the first use of telescopes, Kepler's laws of planetary motion and, of course, the space age.

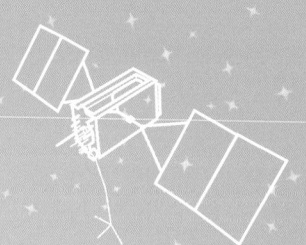

With the dawn of the space era, we no longer needed to look from afar but could send our probes out there to fly past or even land on the object under scrutiny. This has given us amazing insights into what is out there, and has once again evolved our understanding.

With bigger and better telescopes, both Earth-bound and in space, and with every flyby, lander and rover, we get more and more data — so much that gathering, processing and interpreting the information is getting more and more challenging. Teams of scientists across the world are undertaking this process, and due to the sheer volume of data gathered, many scientists are turning to the public with citizen science projects, using our inherent processing capabilities to decipher and gain meaning from what we are finding out in space.

In this book we embrace both ancient and modern data that has been gathered from around our Solar System and have interpreted and displayed it in a stimulating and accessible format, allowing the user quick and easy access to the plethora of information that has been gathered to date.

This book and the graphics contained within used the latest data available at the time of writing. However, data does not stand alone but is interpreted and used to form theories and models. As a result, the graphics generated for this book have needed to illustrate just one of a number of possible explanations and, in some cases, judgment has been used to interpret the available data visually.

As we look to the future, more probes and bigger telescopes are being planned with the hope of getting people out beyond the Moon, to the planets and many other bodies out there in the Solar System. We hope that this book acts as a benchmark for our future forays. Enjoy.

MAGGIE ADERIN-POCOCK

ORIGIN OF THE SOLAR SYSTEM

About 4,600,000,000 years ago, our region of space was home to a "nebula" or cloud of gas and dust. But then …

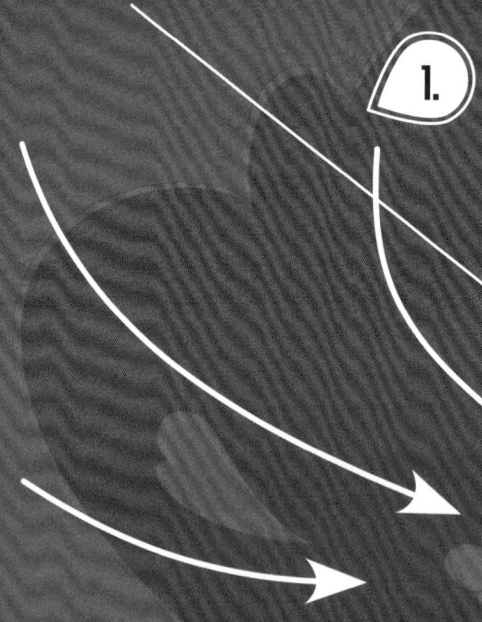

1. The nebula was disturbed by a nearby disruption such as a supernova; the gravitational forces started clumping the gas and dust and pulling it towards the centre.

2. The gas wasn't distributed evenly throughout the nebula, and the pulling towards the centre had a striking effect …

3. The cloud gradually flattened into a disk and began to spin round the centre.

NEBULAR THEORY

This "solar nebular disk model" (SNDM) was supported in 1992 when the Hubble Space Telescope photographed a disk of "protoplanets" forming in the Orion Nebula.

4. Under colossal and increasing pressure, the centre became hotter and hotter.

8. The remaining gas pulled together to form four planets we know as the "gas giants" – Jupiter, Saturn, Uranus and Neptune.

7. Again affected by gravity, the rock and dust in the disk started to form four rocky planets – Mercury, Venus, Earth and Mars – as well as moons, asteroids and other objects...

6. More than 99% of the dust and gas was absorbed by the Sun but some remained.

5. It got so hot and the pressure got so high that atoms of hydrogen fused together, making helium and releasing lots of energy. This was the birth of a star called Sol, better known as the Sun. See **SUN > SOLAR ENERGY** 20

WARNING: NOT TO SCALE

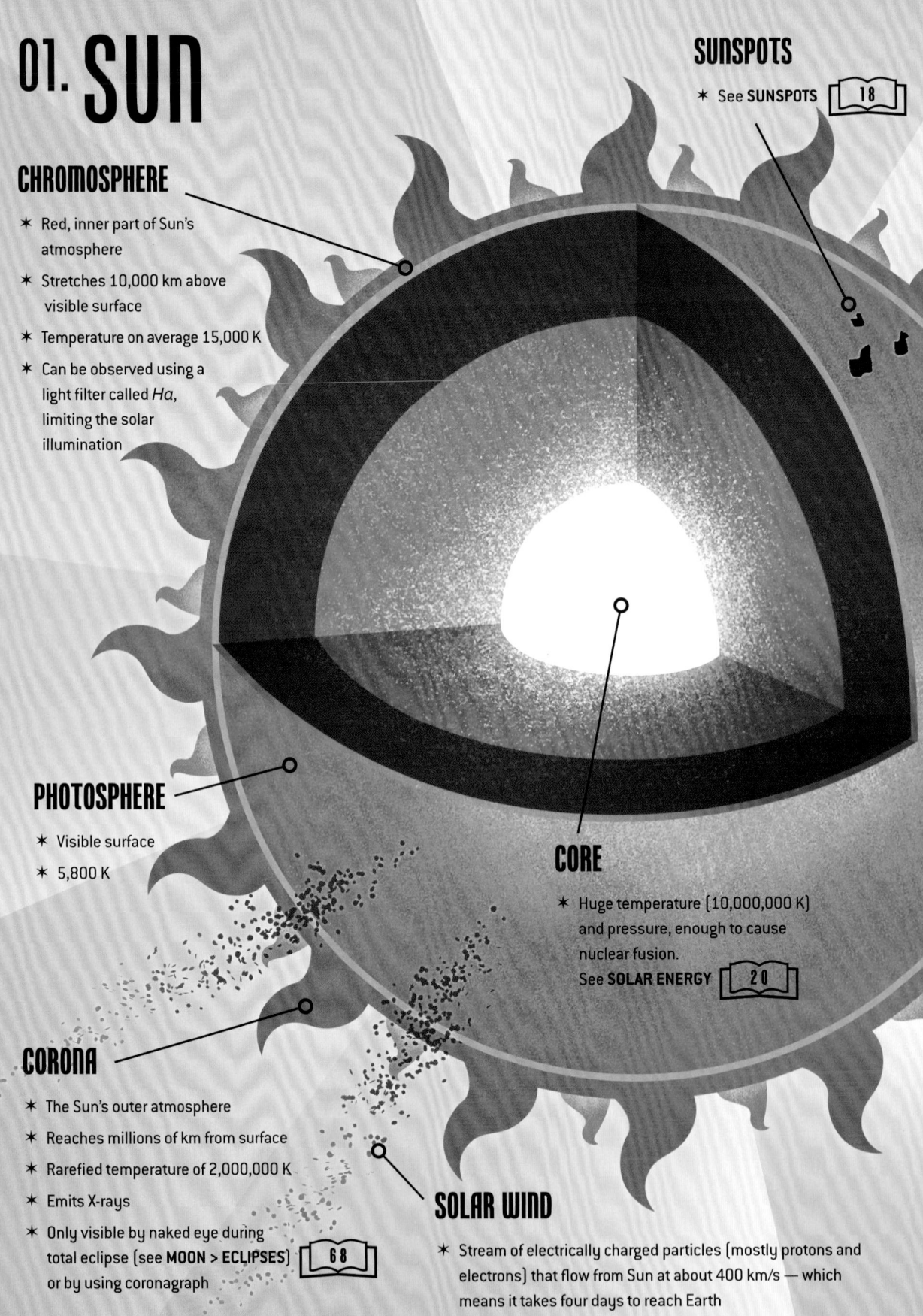

01. SUN

CHROMOSPHERE

* Red, inner part of Sun's atmosphere
* Stretches 10,000 km above visible surface
* Temperature on average 15,000 K
* Can be observed using a light filter called *Ha*, limiting the solar illumination

SUNSPOTS

* See **SUNSPOTS**   [18]

PHOTOSPHERE

* Visible surface
* 5,800 K

CORE

* Huge temperature (10,000,000 K) and pressure, enough to cause nuclear fusion.
See **SOLAR ENERGY**   [20]

CORONA

* The Sun's outer atmosphere
* Reaches millions of km from surface
* Rarefied temperature of 2,000,000 K
* Emits X-rays
* Only visible by naked eye during total eclipse (see **MOON > ECLIPSES**)   [68] or by using coronagraph

SOLAR WIND

* Stream of electrically charged particles (mostly protons and electrons) that flow from Sun at about 400 km/s — which means it takes four days to reach Earth

KEY FACTS

Mass: 330,000 times the Earth
Diameter: 1,400,000 km
Average distance from Earth: 150,000,000 km
Rotation period: 25 Earth days at equator;
36 Earth days at poles

An **astronomical unit (AU)** is a standard measure of distance in space, based on the average distance between the Sun and Earth. 1 AU = 150,000,000 km.

Light from the Sun takes **8 minutes** to reach Earth.

Kelvin (K) is an internationally recognised unit of temperature: 1 K = −273°C and 15,000 K = 14,727°C

BIG STAR

The Sun is huge relative to the eight planets! See below for a size comparison ...

WARNING!
NEVER LOOK DIRECTLY
AT THE SUN!

EARTH   JUPITER

The planets aren't really so close together!

OBSERVING THE SUN

Features of the Sun can be revealed by examining it at different wavelengths of electromagnetic radiation.

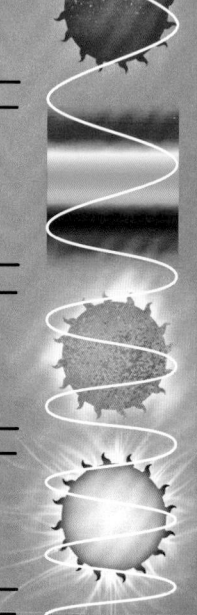

RADIO AND MICROWAVE

Gives blurry image of the Sun, allowing us to see the area where the chromosphere meets the corona

INFRARED

Shows us the chromosphere

VISIBLE LIGHT

Reveals the photosphere

ULTRAVIOLET

Shows us the upper chromosphere/lower corona

X-RAY

Reveals the hottest, most active regions of the sun

GAMMA

Reveals nothing but the most energetic solar flares

JUST MY TYPE

How our Sun compares to the millions of other stars we know of...

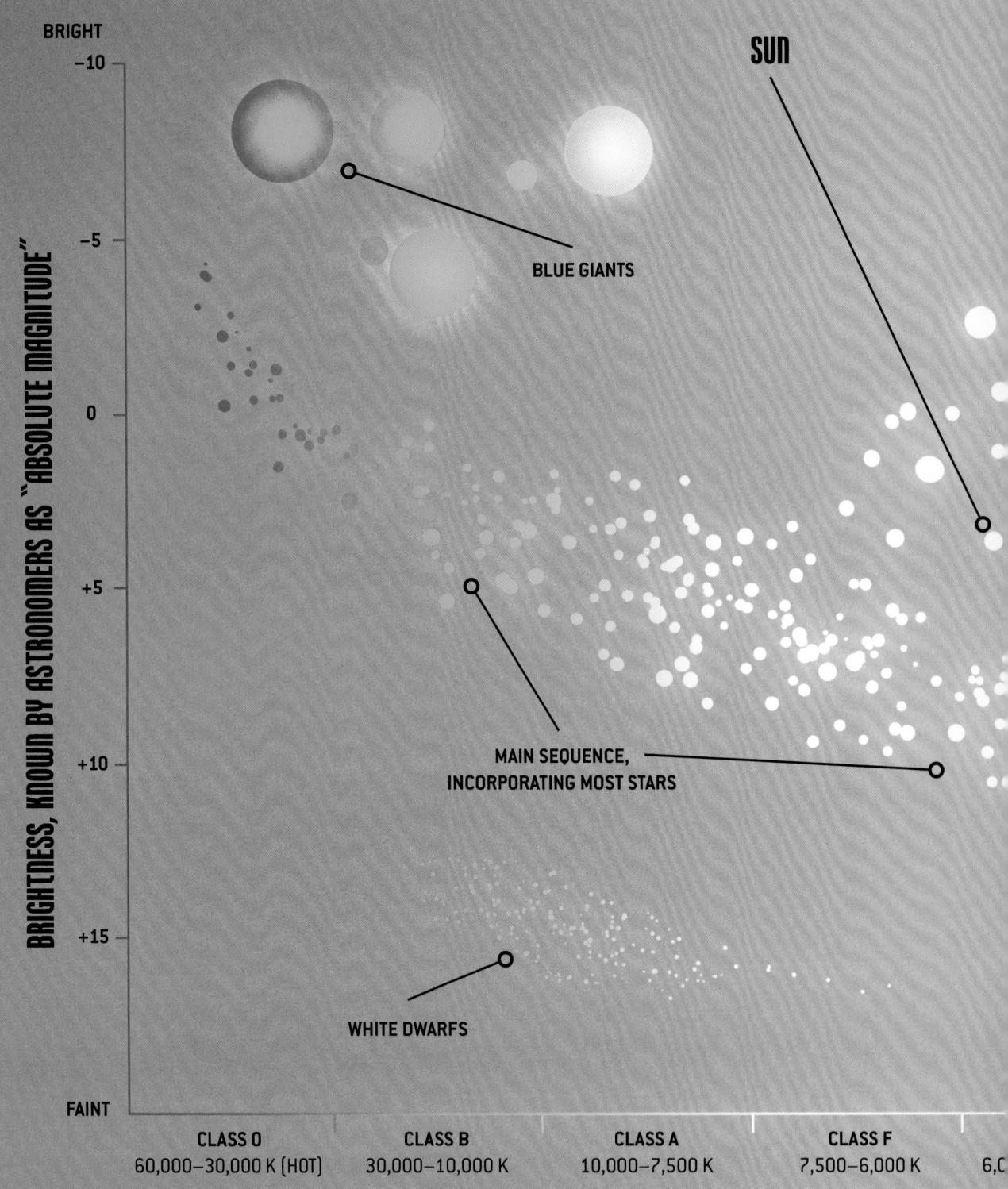

SUN

BRIGHT

−10

−5

0

+5

+10

+15

FAINT

BLUE GIANTS

MAIN SEQUENCE,
INCORPORATING MOST STARS

WHITE DWARFS

BRIGHTNESS, KNOWN BY ASTRONOMERS AS "ABSOLUTE MAGNITUDE"

CLASS O
60,000–30,000 K (HOT)

CLASS B
30,000–10,000 K

CLASS A
10,000–7,500 K

CLASS F
7,500–6,000 K

6,0

SPECTRAL CLASS / TEMPERATURE (K)

RED SUPERGIANTS

RED GIANTS

RED DWARFS

RELATIVE LUMINOSITY (SUN = 1)

- 100,000
- 10,000
- 1,000
- 100
- 10
- 1
- 0.1
- 0.01
- 0.001
- 0.000 1
- 0.000 01

CLASS K	CLASS M
5,000–3,500 K	<3,500 K

0 K

THE FUTURE OF THE SUN

STABLE G2

Our Sun is a stable, yellow G2 main-sequence star, at 4.6 billion years old about midway through its life.

SUN, NOW

RED GIANT

Once the Sun has burned through its fuel for fusion in a few billion years, it will become 250 times bigger than its size today and become a **RED GIANT**. Its diameter of 210 million km will mean that it will consume Mercury, Venus and Earth!

NOT TO SCALE

WHITE DWARF

When the red giant has burned through its fuel, it will then lose layers, forming a planetary nebula (a cloud of gas and plasma) and, eventually, shrink into a **WHITE DWARF** about the same size as Earth, but about 25,000 K!

13

A SENSE OF SCALE

Space is big — really big! Even light from the Sun takes a long time to reach the planets.

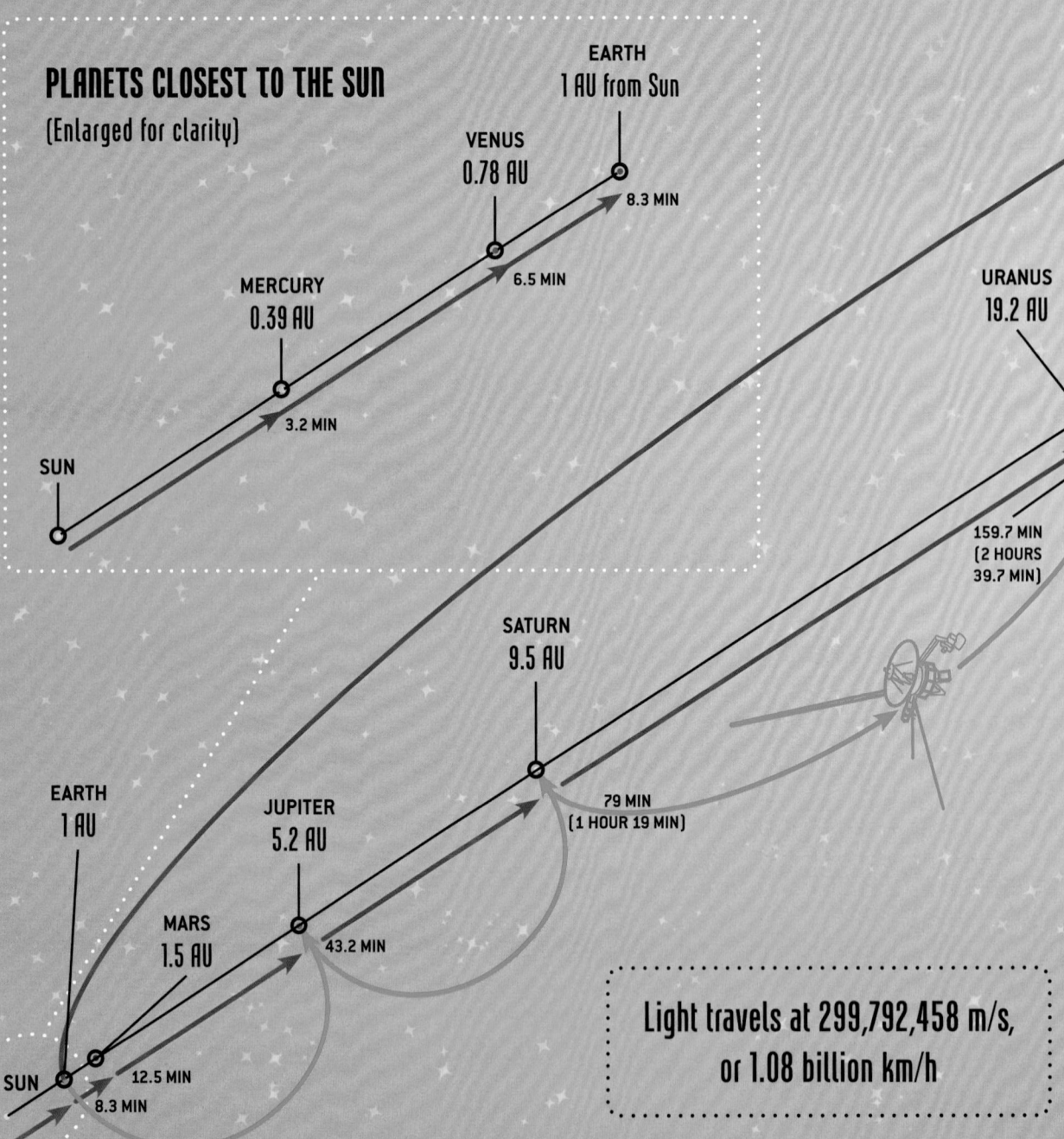

ASTRONOMICAL UNITS (AU)

The Solar System is so big we measure distance using a larger scale than kilometres (km), based on the average distance from Earth and to the Sun: 1 AU = 150 million km.

PLANETS CLOSEST TO THE SUN
(Enlarged for clarity)

EARTH
1 AU from Sun
8.3 MIN

VENUS
0.78 AU
6.5 MIN

MERCURY
0.39 AU
3.2 MIN

SUN

URANUS
19.2 AU

159.7 MIN
(2 HOURS
39.7 MIN)

SATURN
9.5 AU
79 MIN
(1 HOUR 19 MIN)

EARTH
1 AU

JUPITER
5.2 AU
43.2 MIN

MARS
1.5 AU
12.5 MIN

SUN
8.3 MIN

Light travels at 299,792,458 m/s, or 1.08 billion km/h

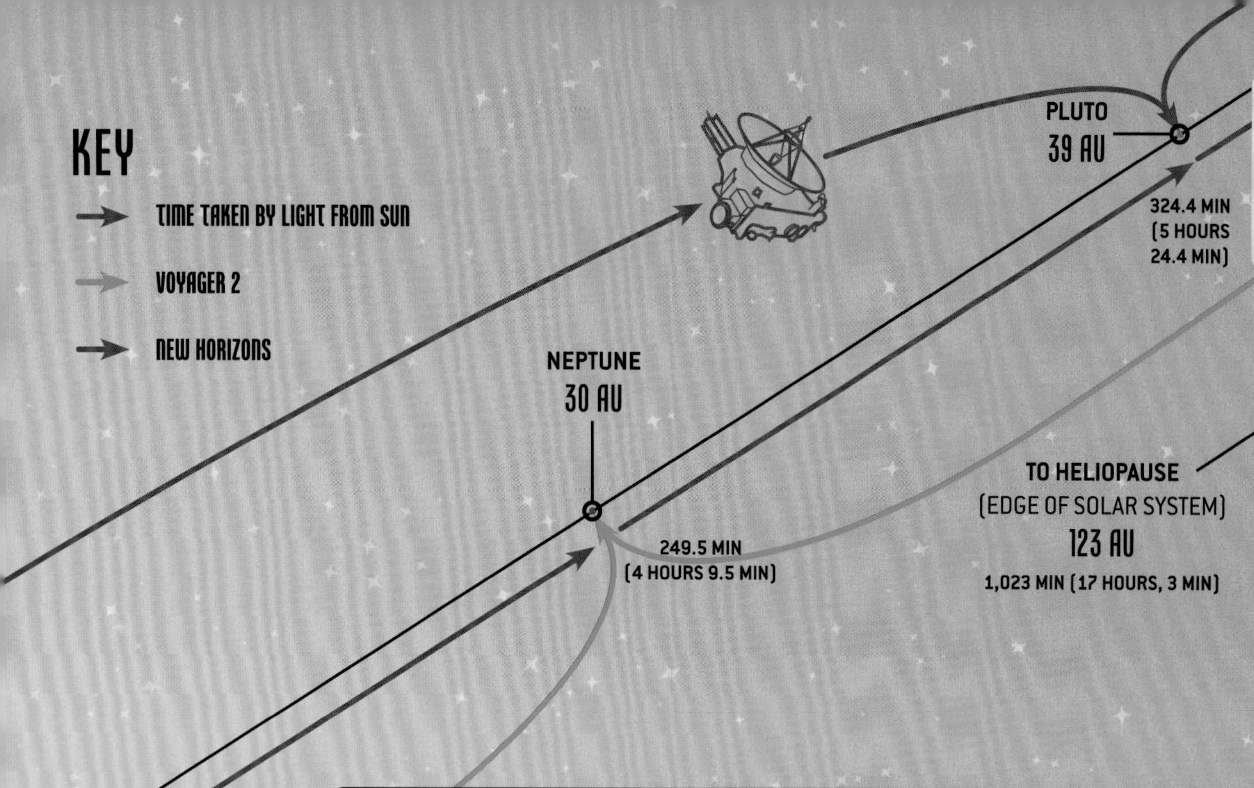

KEY

→ TIME TAKEN BY LIGHT FROM SUN

→ VOYAGER 2

→ NEW HORIZONS

PLUTO
39 AU

324.4 MIN
(5 HOURS
24.4 MIN)

NEPTUNE
30 AU

249.5 MIN
(4 HOURS 9.5 MIN)

TO HELIOPAUSE
(EDGE OF SOLAR SYSTEM)
123 AU

1,023 MIN (17 HOURS, 3 MIN)

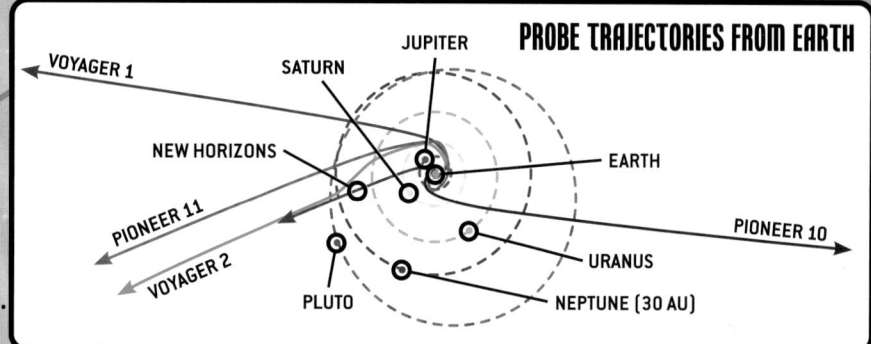

PROBE TRAJECTORIES FROM EARTH

VOYAGER 1
SATURN
JUPITER
NEW HORIZONS
EARTH
PIONEER 11
PIONEER 10
VOYAGER 2
URANUS
PLUTO
NEPTUNE (30 AU)

OUR FURTHEST PROBES

VOYAGER 1 [still active, intermittently] →

* **5 September 1977** — Launched from Earth

* **25 August 2012** — Passed heliopause, first human-made object to leave Solar System

* **5 January 2024** — At 162.9 AU from Earth, moving at 60,970 km/h

VOYAGER 2 [still active] →

* **20 August 1977** — Launched from Earth

* **5 January 2024** — At 136.2 AU, moving at 55,066 km/h

PIONEER 10 →

* **3 March 1972** — Launched from Earth

* **23 January 2003** — Last communication received at 80 AU

* **5 January 2024** — At 134.2 AU

PIONEER 11 →

* **6 April 1973** — Launched from Earth

* **24 November 1995** — Last communication received at 44.7 AU

* **5 January 2024** — At 112.5 AU

NEW HORIZONS [still active] →

* **19 January 2006** — Launched from Earth

* **5 January 2024** — 58.9 AU, moving at 49,320 km/h

THE WAY YOU MOVE

The movements of the Sun, planets and other stars as seen from Earth.

CELESTIAL SPHERE

An imaginary sphere in which Earth sits at the centre, and onto which all the celestial bodies of the night sky can be projected. The celestial poles and equator extend from the poles and equator on Earth.

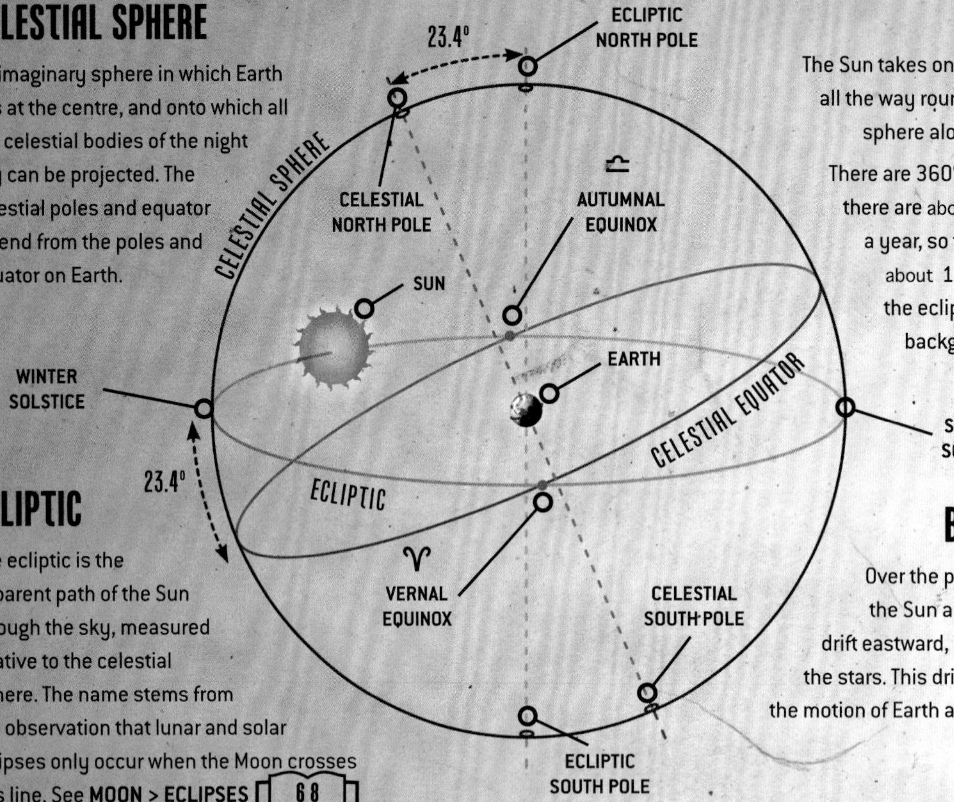

ECLIPTIC

The ecliptic is the apparent path of the Sun through the sky, measured relative to the celestial sphere. The name stems from the observation that lunar and solar eclipses only occur when the Moon crosses this line. See **MOON > ECLIPSES** `6 8`

CELESTIAL SPHERE

23.4°

ECLIPTIC NORTH POLE

CELESTIAL NORTH POLE

AUTUMNAL EQUINOX

SUN

EARTH

WINTER SOLSTICE

CELESTIAL EQUATOR

23.4°

ECLIPTIC

SUMMER SOLSTICE

VERNAL EQUINOX

CELESTIAL SOUTH POLE

ECLIPTIC SOUTH POLE

SUN PATH

The Sun takes one year to move all the way round the celestial sphere along the ecliptic.

There are 360° in a circle and there are about 360 days in a year, so the Sun moves about 1° per day along the ecliptic against the background of stars.

BACK-SPIN

Over the period of a year, the Sun appears to us to drift eastward, lagging behind the stars. This drift is caused by the motion of Earth around the Sun.

VERNAL EQUINOX

An equinox occurs when the Sun passes directly over the equator and day length in the northern and southern hemispheres are equal. There are two equinoxes each year: the spring or vernal equinox occurs in March, and the autumnal equinox occurs in September. The vernal equinox is used as the zero point of longitude on the celestial sphere coordinate system. If the ecliptic is plotted on the grid of a star chart using celestial coordinates, it appears as a curved line (due to Earth's "obliquity" or tilt to the plane of our orbital path around the Sun).

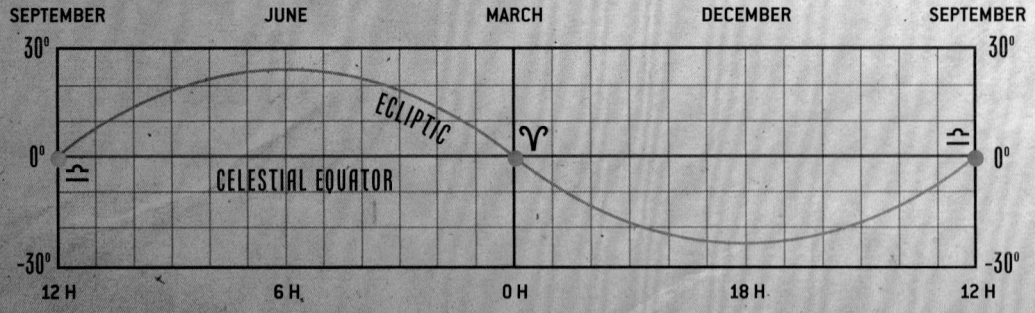

WINTER AND SUMMER AS VIEWED FROM THE NORTHERN HEMISPHERE

16

ZODIAC

The other seven planets of the Solar System orbit the Sun on roughly the same plane as the Earth. This means that, from Earth, they always appear close to the ecliptic — within 8° above or below it.

We call this 16° band around the ecliptic in which the planets appear the **ZODIAC**.

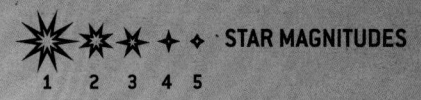
STAR MAGNITUDES
1 2 3 4 5

THE VERNAL EQUINOX
CURRENTLY OCCURS IN PISCES

PISCES (PSC)
AQUARIUS (AQR)
CAPRICORNUS (CAP)
ARIES (ARI)
SAGITTARIUS (SGR)
TAURUS (TAU)

MARCH 21
FEBRUARY 21
APRIL 21
JANUARY 21
MAY 21
OCTOBER 21 SEPTEMBER 21 AUGUST 21
JULY 21
NOVEMBER 21
DECEMBER 21
JUNE 21
DECEMBER 21
EARTH'S POSITION IN ORBIT
MAY 21
JUNE 21
JANUARY 21 FEBRUARY 21 MARCH 21 APRIL 21
NOVEMBER 21
THE SUN'S APPARENT POSITION AS SEEN FROM EARTH
JULY 21
OCTOBER 21
AUGUST 21 SEPTEMBER 21

GEMINI (GEM)
VIRGO (VIR)
OPHIUCHUS (OPH)
LEO (LEO)
LIBRA (LIB)
CANCER (CNC)
SCORPIUS (SCO)

THIRTEEN CONSTELLATIONS

The astrological zodiac is often divided into 12 signs, the dates of which are fixed relative to the equinox. Astronomers divide the zodiac into a slightly different system: 13 constellations of stars that the Sun is seen to pass through over the course of a year.

"TRAVELLER"

In ancient times, the word "planet" or "traveller" meant objects in the night sky that did not move in the same way as stars. We now know that the "wandering" movement of planets is because they orbit the Sun.

SUNSPOTS

Charting the storms on the surface of the Sun.

INTRODUCING SUNSPOTS

Sunspots are the most noticeable feature on the surface of the Sun.

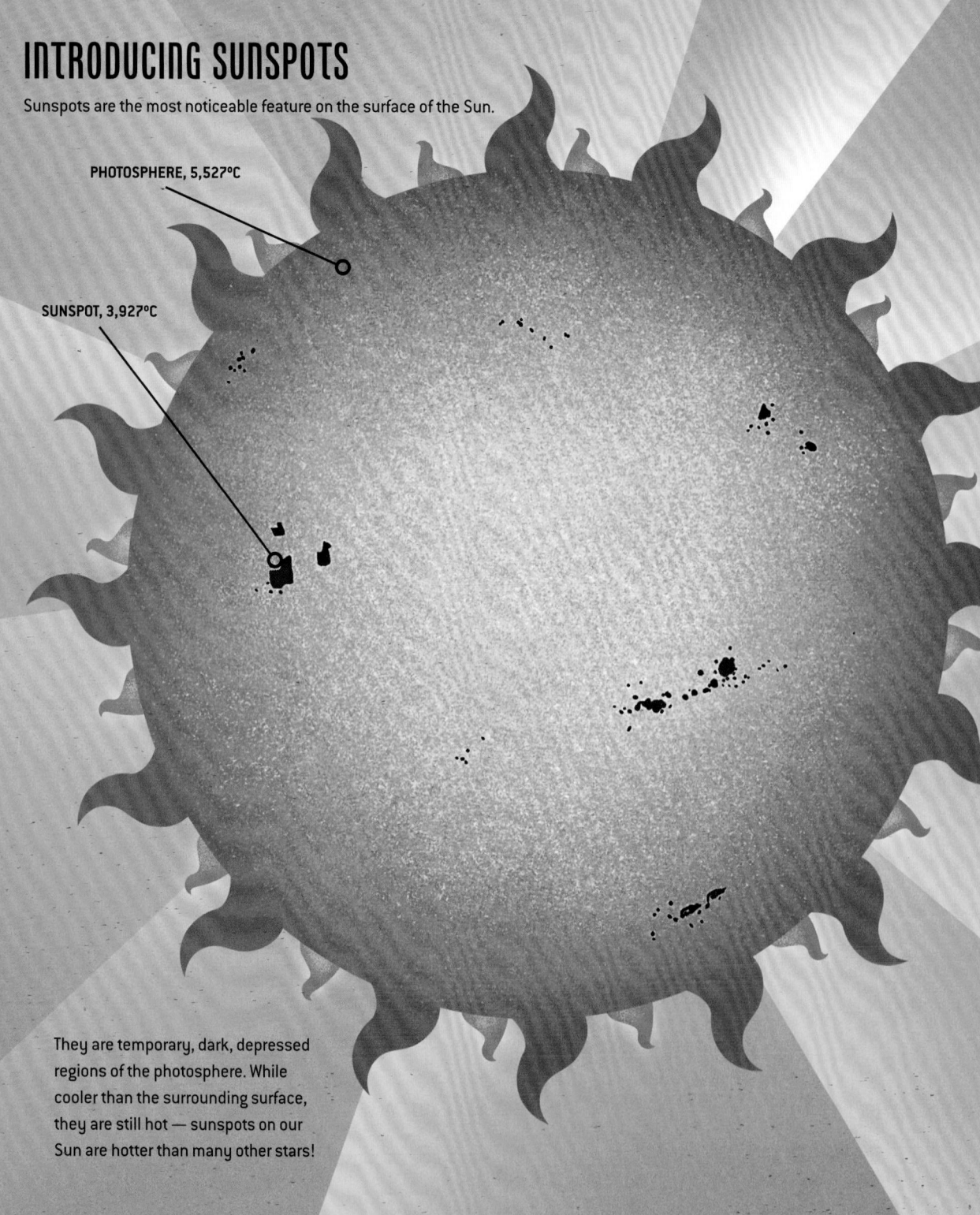

PHOTOSPHERE, 5,527°C

SUNSPOT, 3,927°C

They are temporary, dark, depressed regions of the photosphere. While cooler than the surrounding surface, they are still hot — sunspots on our Sun are hotter than many other stars!

WHY ARE THERE SUNSPOTS?

The Sun rotating faster at its equator than at its poles means its magnetic fields get twisted, preventing the convection of heat to the surface.

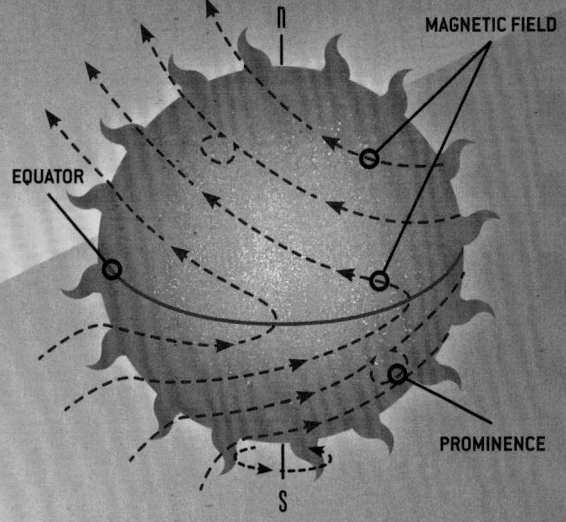

N

MAGNETIC FIELD

EQUATOR

PROMINENCE

S

USING SUNSPOTS

By watching sunspots cross the Sun's surface, we can measure the speed of the Sun's rotation at different latitudes:

* At poles = 36 Earth days

* At equator = 25 Earth days

Understanding sunspot activity also helps us predict solar activity and solar storms.

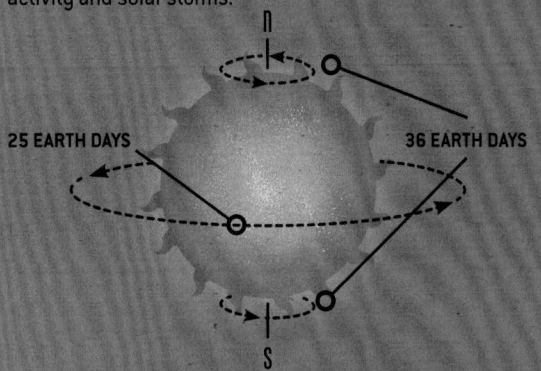

N

25 EARTH DAYS

36 EARTH DAYS

S

HOW LONG DO SUNSPOTS LAST?

* Individual sunspots or sunspot groups grow and fade over days or weeks

* Individual sunspots can be seen from Earth for no more than two weeks due to Sun's rotation

"BUTTERFLY" DIAGRAM

By plotting sunspots on a graph, we see a regular cycle of solar activity lasting about 11 years.

DAILY SUNSPOT AREA AVERAGE OVER INDIVIDUAL SOLAR ROTATIONS

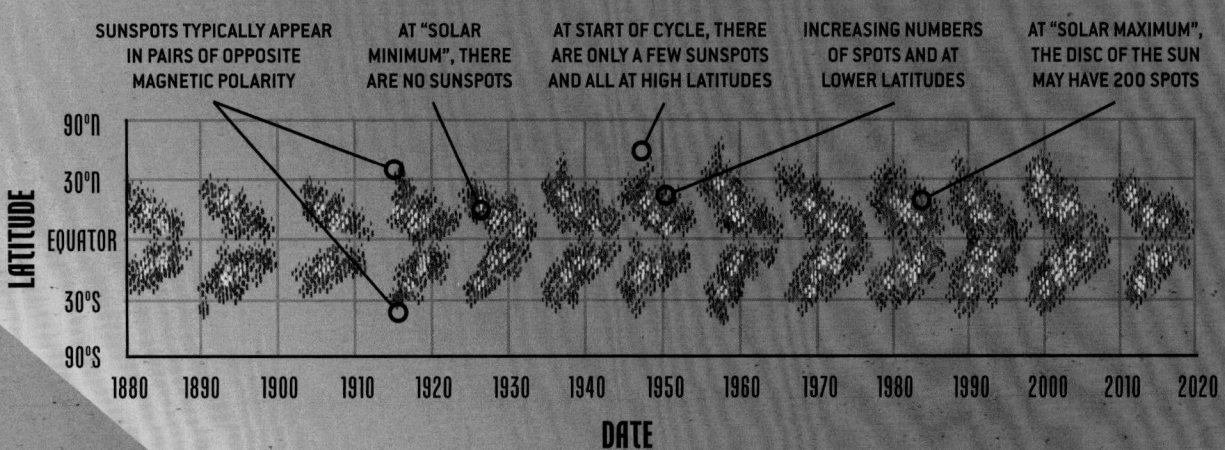

SUNSPOTS TYPICALLY APPEAR IN PAIRS OF OPPOSITE MAGNETIC POLARITY

AT "SOLAR MINIMUM", THERE ARE NO SUNSPOTS

AT START OF CYCLE, THERE ARE ONLY A FEW SUNSPOTS AND ALL AT HIGH LATITUDES

INCREASING NUMBERS OF SPOTS AND AT LOWER LATITUDES

AT "SOLAR MAXIMUM", THE DISC OF THE SUN MAY HAVE 200 SPOTS

LATITUDE

90°N · 30°N · EQUATOR · 30°S · 90°S

1880 1890 1900 1910 1920 1930 1940 1950 1960 1970 1980 1990 2000 2010 2020

DATE

At solar maximum, the Sun's surface is generally more active, too — with more solar flares and prominences. This can affect electrical systems on Earth, so it's good to be able to predict such solar weather!

SOLAR ENERGY

The light and heat radiated by the Sun are the result of a process called nuclear fusion.

NUCLEAR FUSION

E=mc² — Albert Einstein's famous equation explains the huge amount of energy released from a given amount of mass: energy = mass x the speed of light squared.

The enormous pressure and temperature at the centre of the Sun enables Hydrogen atoms to fuse together, releasing huge amounts of energy in the process.

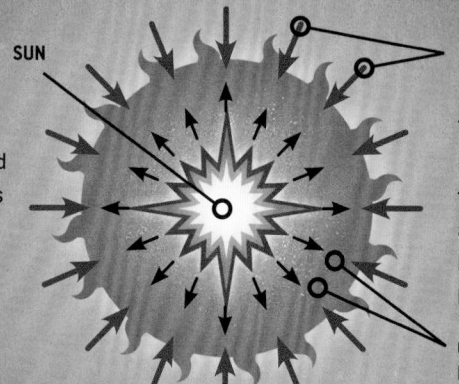

SUN

INWARD PRESSURE FROM GRAVITY

This energy keeps the Sun burning very brightly — even though it is 150 million km away from Earth, the Sun is so bright we can damage our eyes by looking at it directly.

OUTWARD PRESSURE FROM NUCLEAR FUSION

NUCLEAR FUSION IN THE SUN

STAGE 1: Average atom of ¹H in the core of the Sun waits 9 million years before fusing

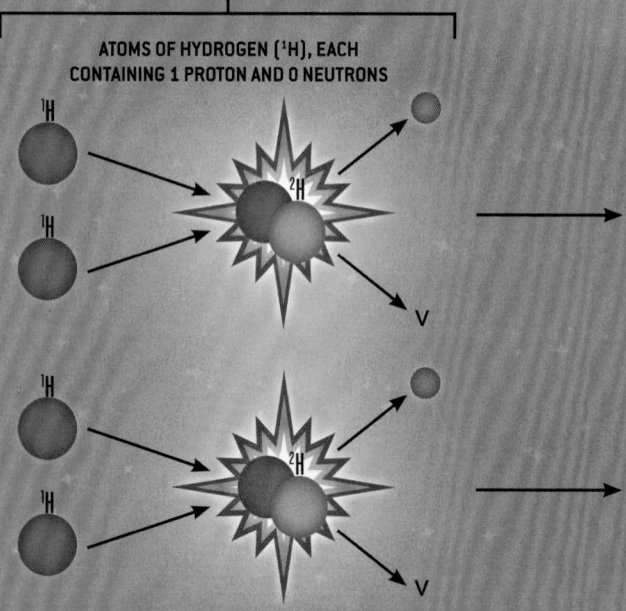

ATOMS OF HYDROGEN (¹H), EACH CONTAINING 1 PROTON AND 0 NEUTRONS

Under huge pressure, two ¹H atoms fuse together.

* One of the protons undergoes "beta plus decay", converting into a neutron by releasing parts of itself in the form of 1 positron and 1 electron neutrino.

* The result is 1 atom of deuterium (²H) — a type of hydrogen (or "isotope") containing 1 proton and 1 neutron.

* The released positron then usually collides with an electron and annihilates — releasing energy.

STAGE 2: Average atom of ²H in the core of the Sun waits 4 seconds to fuse with an atom of ¹H

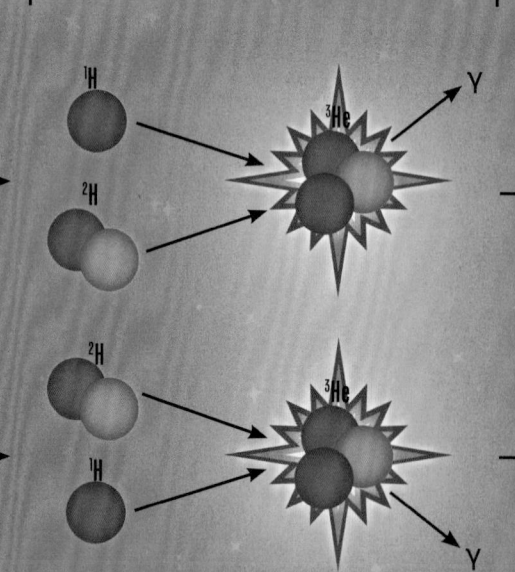

Still under pressure, one 2H atom fuses with an atom of ¹H.

* The result is 1 atom of Helium-3 (³He), an isotope of Helium containing 2 protons and 1 neutron.

* A gamma ray and energy are also released.

HOW THE SUN'S ENERGY FUELS US

SUN

Humans convert direct sunlight into vitamin D to keep bones, teeth and muscles healthy.

Plants use photosynthesis to convert sunlight (plus carbon dioxide and water) into glucose and oxygen.

Humans breathe oxygen and eat plants, converting sugars to fat and protein.

Humans eat animals and animal products, absorbing fats and protein.

Animals breathe oxygen and eat plants, converting sugars to fat and protein.

Fossil fuels such as petroleum, coal and natural gas are produced by decomposing life forms, so are also made up of energy from the Sun.

STAGE 3: Average atom of ^{3}He waits 400 years before fusing to become ^{4}He.

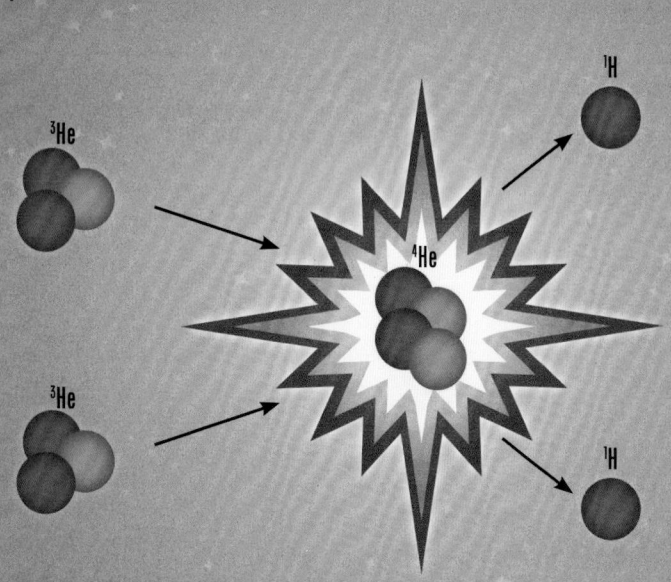

Still under pressure, 2 atoms of ^{3}He fuse together.

* The result is 1 atom of Helium-4 (^{4}He), containing 2 protons and 2 neutrons.
* In 83.3% of times, the result is also 2 atoms of ^{1}H and energy.
* In 16.7% of times, the result is also ^{1}H, isotopes of Beryllium and Lithium, and energy.

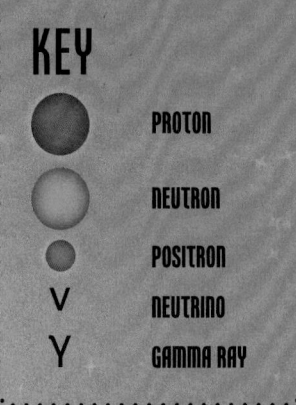

KEY

- PROTON
- NEUTRON
- POSITRON
- v NEUTRINO
- γ GAMMA RAY

HOW MUCH ENERGY IS RELEASED?

Converting the mass contained in a glass of water could power an entire city for a year!

02. MERCURY

CRUST

* Heavily cratered
* Largely low iron basalt

ATMOSPHERE

* Mercury is small, so has weak gravitational field and thin atmosphere
* The atmosphere may once have been thicker, but the planet's low mass means it may have been eroded by solar wind

MANTLE

* 600 km thick

CRUST = 100–200 KM

CORE

* Large, partly fluid iron core 3,600 km in diameter (75% of total!)

SURFACE

* 40% = smooth plains that resemble maria (the dark "seas" on Earth's Moon) but reflect more sunlight so are brighter
* About 45% = ancient, heavily cratered and degraded plains
* 15% = giant impact basins and ejecta

SCARP

* A long, cliff-like face or slope
* Lobate thrust scarps formed where the surface crust compressed as it cooled

KEY FACTS

Diameter: 4,900 km
Average distance from Sun: 58 million km
Average distance from Earth: 77 million km
Day: 58 Earth days, 15 hours and 30 minutes
Year: 88 Earth days
Moons: 0

HOT AND COLD

SUN-BLASTED EQUATOR:	POLAR CRATERS IN ALMOST PERMANENT SHADOW:
450°C	-180°C

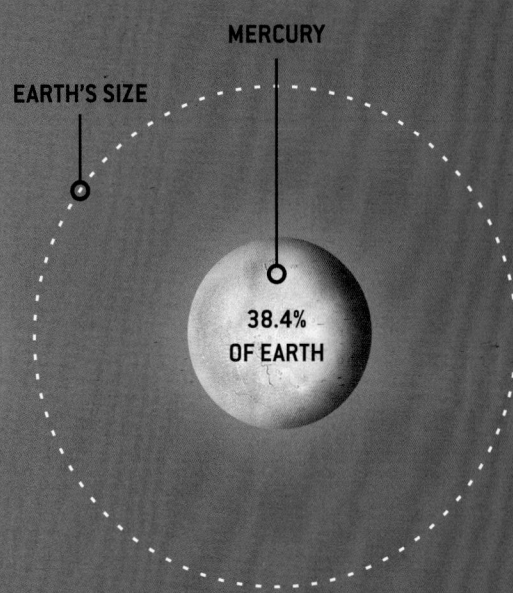

MERCURY

SUN

SCALE

EARTH

MERCURY

EARTH'S SIZE

38.4% OF EARTH

ICE SPY

Radar mapping suggests there is water ice in the permanently shadowed polar craters.

KEY

 WATER ICE

 SHADOW

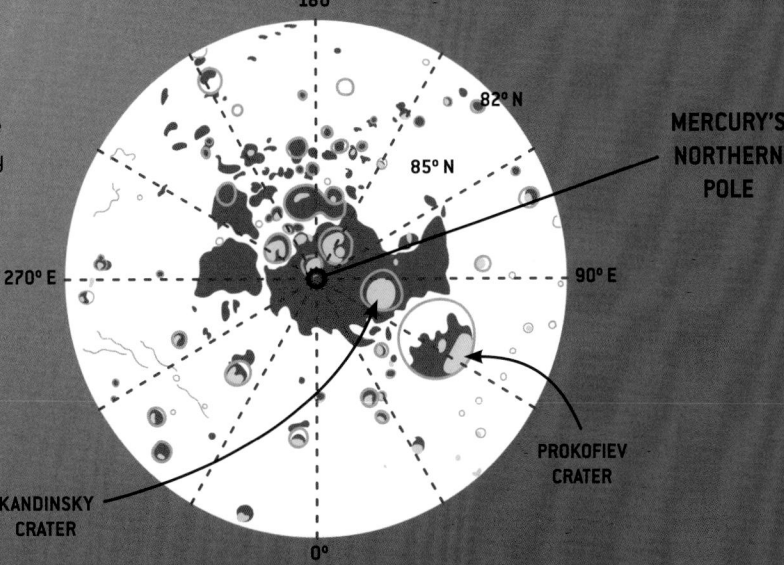

180°

82° N

85° N

270° E

90° E

MERCURY'S NORTHERN POLE

0°

KANDINSKY CRATER

PROKOFIEV CRATER

SPACE JARGON – I

The technical language used by astronomers — made easy.

INFERIOR PLANET

Mercury and Venus, which orbit closer to the Sun than Earth, are known as inferior planets.

To read about Superior Planets, see
MARS > SPACE JARGON – II

📖 92

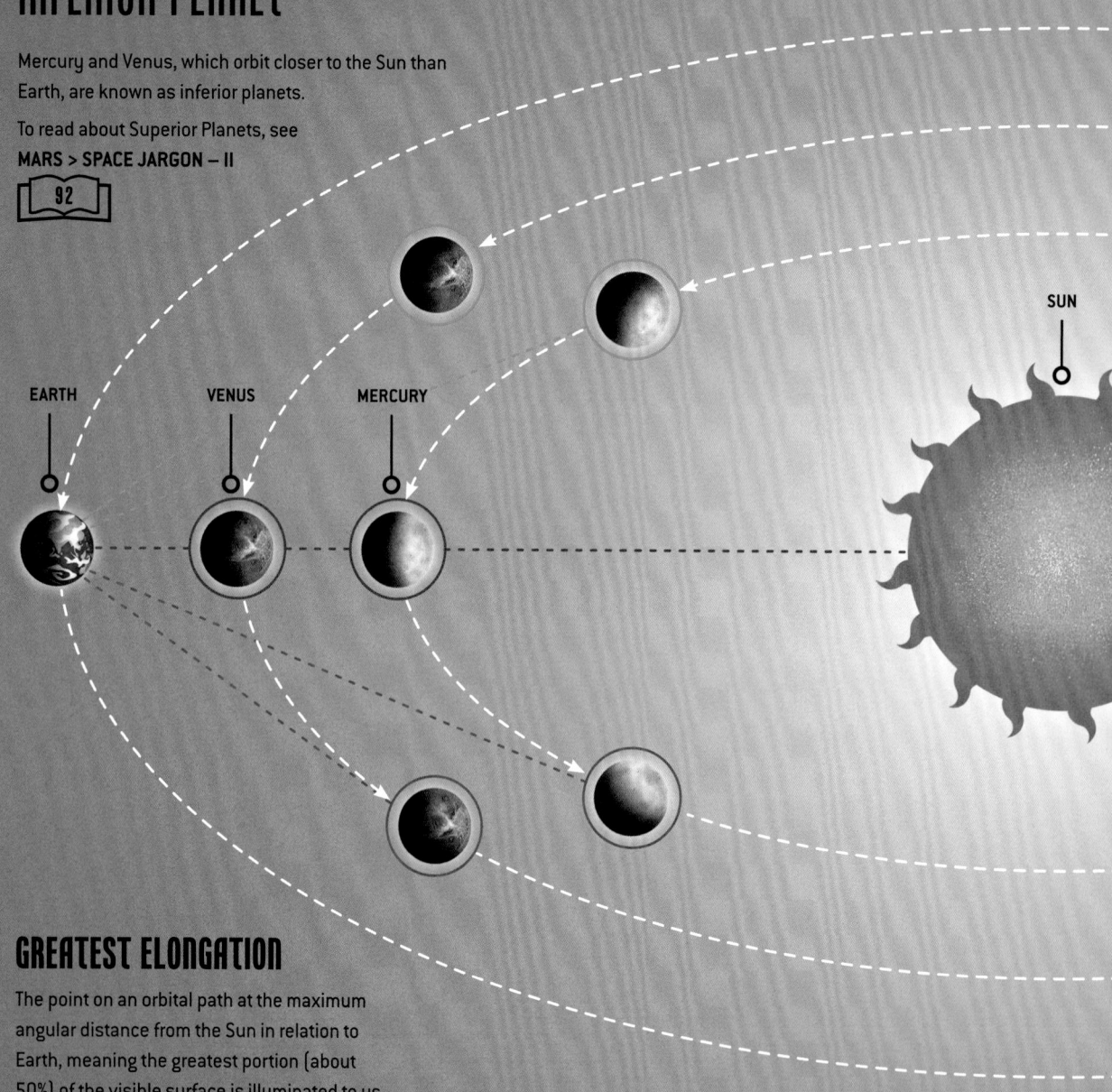

SUN

EARTH

VENUS

MERCURY

GREATEST ELONGATION

The point on an orbital path at the maximum angular distance from the Sun in relation to Earth, meaning the greatest portion (about 50%) of the visible surface is illuminated to us

GREATEST WESTERN ELONGATION

Mercury and Venus are west of the Sun so visible with the naked eye from Earth just after sunrise

GREATEST EASTERN ELONGATION

Mercury and Venus are east of the Sun so visible with the naked eye from Earth just after sunset

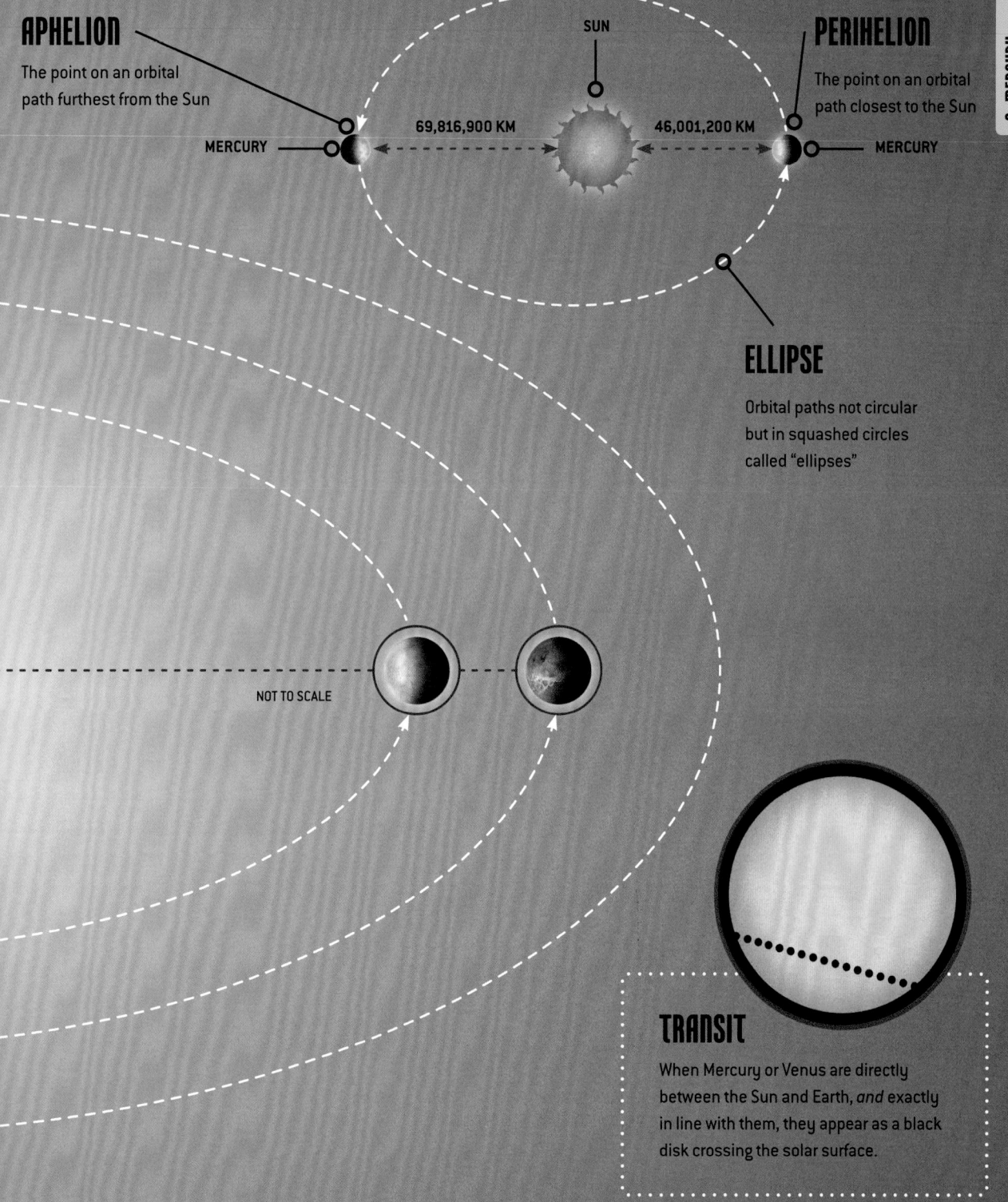

APHELION

The point on an orbital
path furthest from the Sun

SUN

69,816,900 KM

46,001,200 KM

PERIHELION

The point on an orbital
path closest to the Sun

MERCURY

MERCURY

ELLIPSE

Orbital paths not circular
but in squashed circles
called "ellipses"

NOT TO SCALE

TRANSIT

When Mercury or Venus are directly
between the Sun and Earth, *and* exactly
in line with them, they appear as a black
disk crossing the solar surface.

INFERIOR CONJUNCTION

When Mercury or Venus are directly between the Earth and
Sun (invisible except in a **TRANSIT**, see above right)

SUPERIOR CONJUNCTION

When an inferior planet is on the far side of the Sun

WHAT IS A PLANET?

In 2006, the International Astronomical Union defined a planet with three criteria.

A PLANET IS A CELESTIAL BODY THAT:

1. Orbits a star*.

* For example, the Sun.

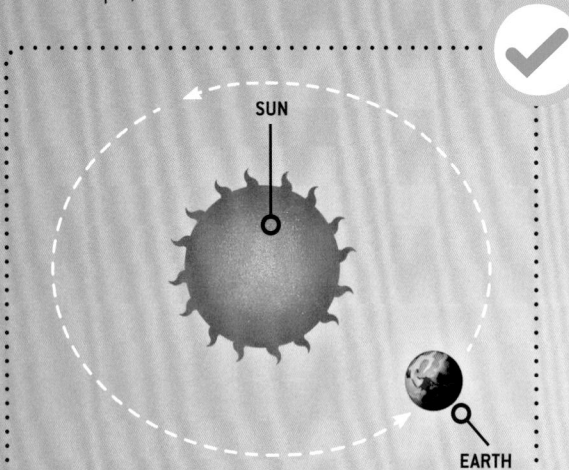

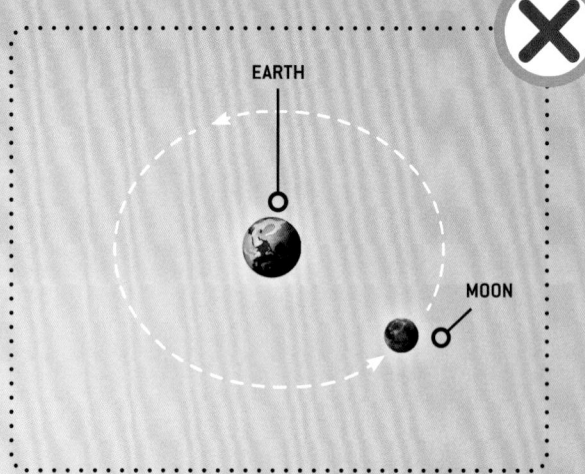

AND

2. Is massive enough for its own gravity to make it (nearly) spherical.

AND

3. Has cleared the neighbourhood round its orbit.

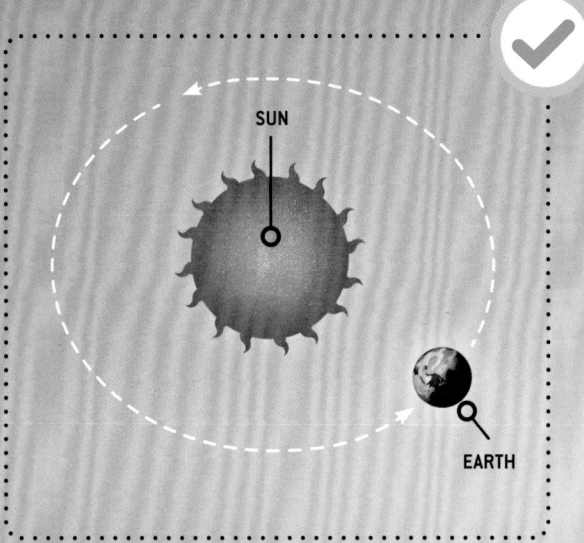

SUN

EARTH

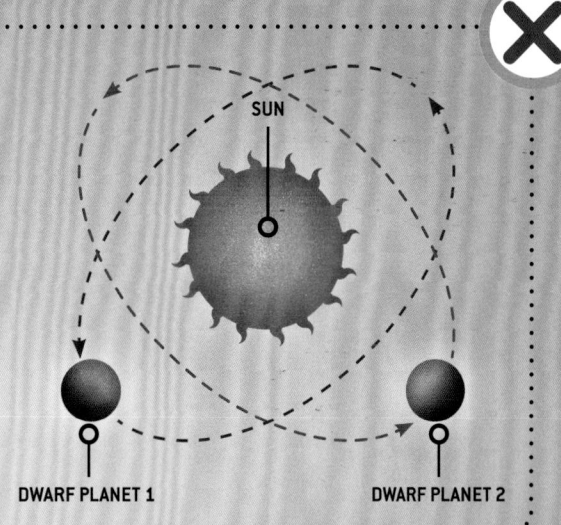

SUN

DWARF PLANET 1

DWARF PLANET 2

DWARF PLANETS

A **DWARF PLANET** is a celestial body that (1) orbits a star and (2) is nearly spherical but (3) has NOT cleared the neighbourhood round its orbit. **

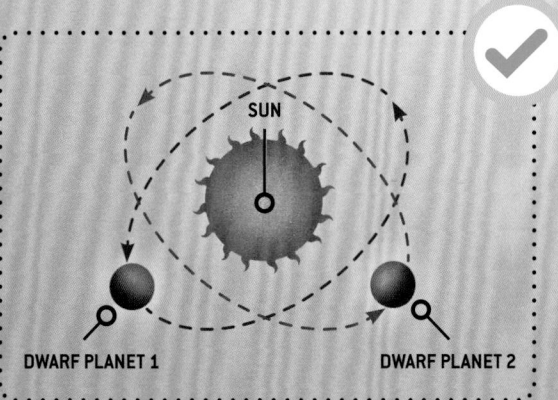

SUN

DWARF PLANET 1

DWARF PLANET 2

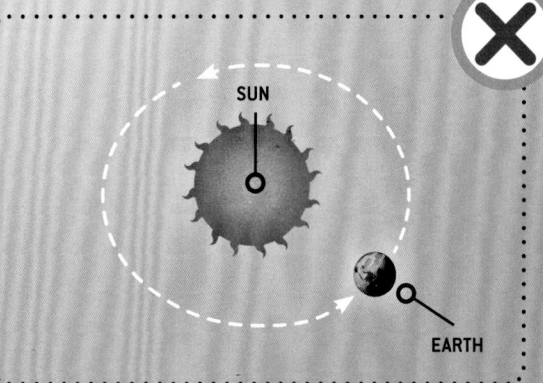

SUN

EARTH

** And is not a satellite or moon.

SMALL SOLAR-SYSTEM BODIES

All other objects — except satellites (moons) — that orbit the Sun are defined as **SMALL SOLAR-SYSTEM BODIES**.

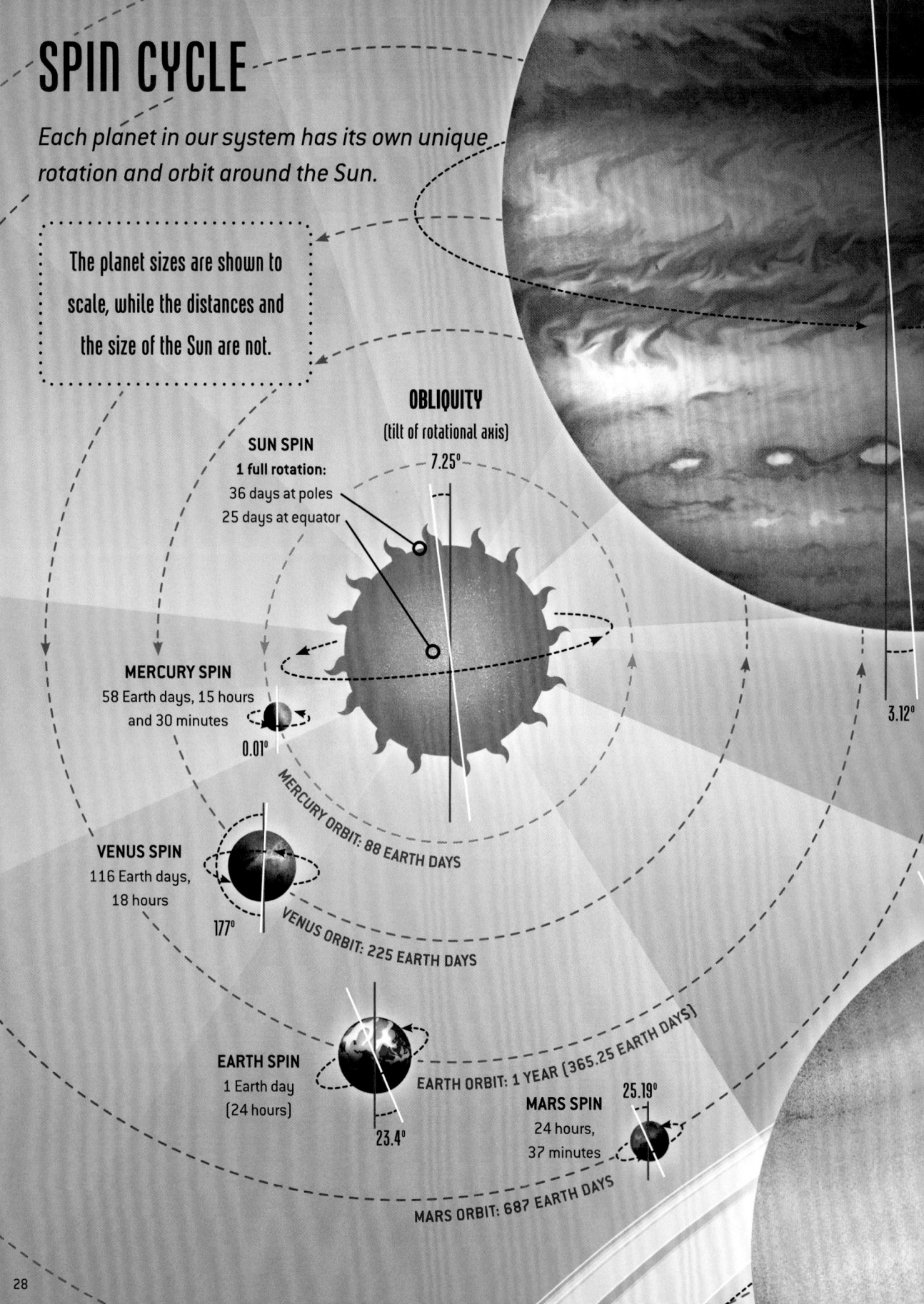

SPIN CYCLE

Each planet in our system has its own unique rotation and orbit around the Sun.

The planet sizes are shown to scale, while the distances and the size of the Sun are not.

OBLIQUITY
(tilt of rotational axis)

7.25°

SUN SPIN
1 full rotation:
36 days at poles
25 days at equator

3.12°

MERCURY SPIN
58 Earth days, 15 hours and 30 minutes

0.01°

MERCURY ORBIT: 88 EARTH DAYS

VENUS SPIN
116 Earth days, 18 hours

177°

VENUS ORBIT: 225 EARTH DAYS

EARTH SPIN
1 Earth day
(24 hours)

23.4°

EARTH ORBIT: 1 YEAR (365.25 EARTH DAYS)

25.19°

MARS SPIN
24 hours, 37 minutes

MARS ORBIT: 687 EARTH DAYS

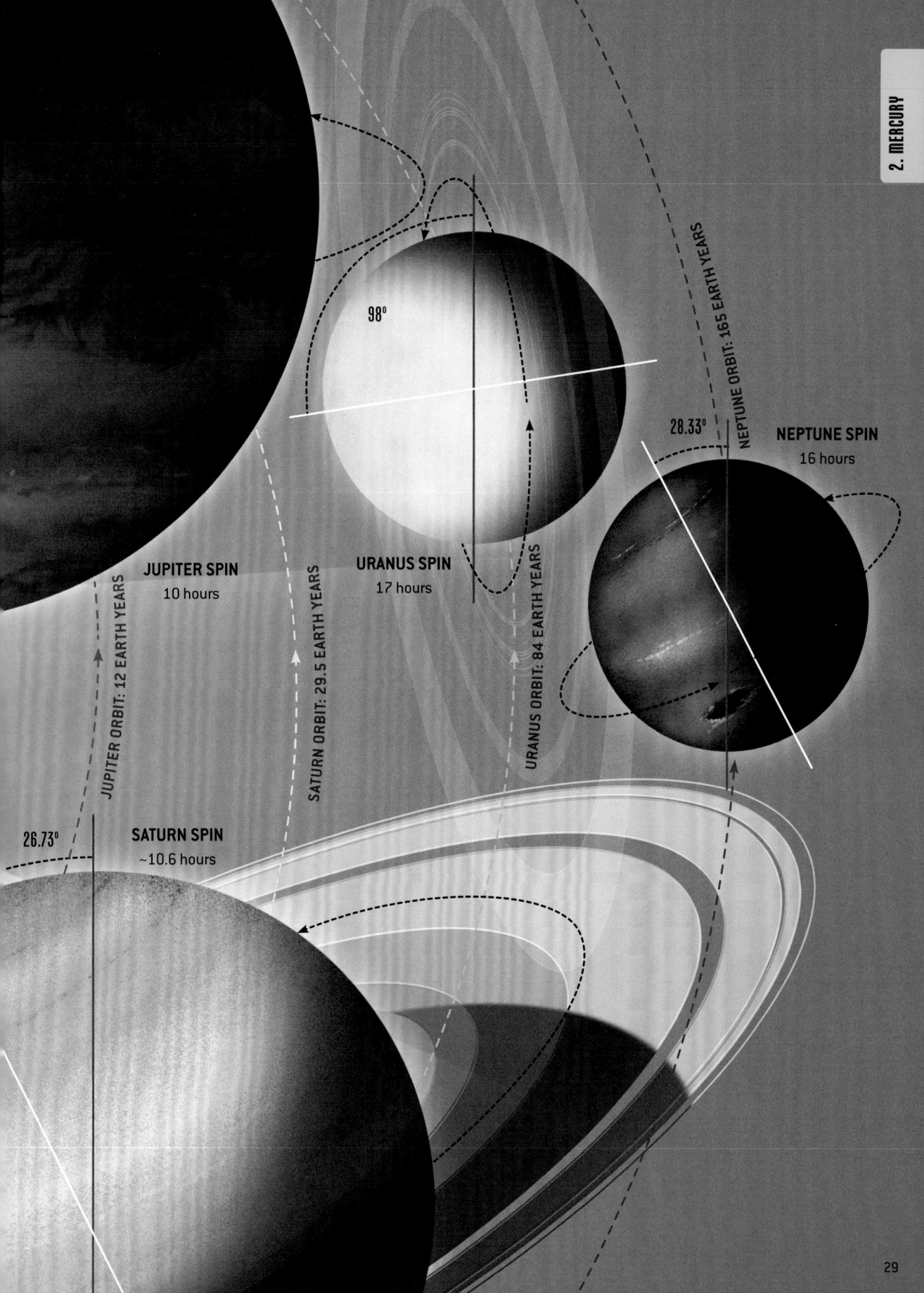

98°

NEPTUNE SPIN
16 hours

28.33°

NEPTUNE ORBIT: 165 EARTH YEARS

JUPITER SPIN
10 hours

URANUS SPIN
17 hours

JUPITER ORBIT: 12 EARTH YEARS

SATURN ORBIT: 29.5 EARTH YEARS

URANUS ORBIT: 84 EARTH YEARS

26.73°

SATURN SPIN
~10.6 hours

PROBES – I

The craft we've sent to fly past or orbit the Sun and inner planets.

SUN — 21

1960 – Pioneer 5
1965 – Pioneer 6
1966 – Pioneer 7
1967 – Pioneer 8
1968 – Pioneer 9
1974 – Helios A
1976 – Helios B
1978 – ISEE-3
1994 – Ulysses
1994 – WIND
1996 – SOHO
1997 – ACE
2001 – Genesis
2006 – STEREO A
2006 – STEREO B
2015 – DSCOVR
2018 – Parker Solar Probe
2020 – Solar Orbiter
2022 – ASO-S
2022 – CuSP
2023 – Aditya-L1

MERCURY — 3

1974 – Mariner 10
2008 – MESSENGER [finally crashed on surface 2015]
2018 – BepiColombo

*VENUSIAN PROBES

These probes landed or crashed on the surface. See **VENUS > EXPLORERS**

38

SUN, FAILED — 1

1969 – Pioneer E

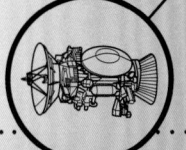

CASSINI-HUYGENS

2017 – crashed into Saturn

See **JUPITER > GRAVITATIONAL SLINGSHOT**

110

ULYSSES

VENUS — 29

1962 – Mariner 2
1967 – Venera 4*
1967 – Mariner 5
1969 – Venera 5*
1969 – Venera 6*
1979 – Venera 7*
1972 – Venera 8*
1974 – Mariner 10 [on way to Mercury]
1975 – Venera 9*
1975 – Venera 10*
1978 – Pioneer Venus*
1978 – Venera 12*
1978 – Venera 11*
1982 – Venera 13*
1982 – Venera 14*
1983 – Venera 15
1983 – Venera 16
1985 – Vega 1*
1985 – Vega 2*
1990 – Galileo
1990 – Magellan
1998 – Cassini [gravity assist on way to Saturn]
2006 – Venus Express
2006 – MESSENGER [on way to Mercury]
2010 – IKAROS
2015 – Akatsuki
2018 – Parker Solar Probe [on way to Sun]
2018 – BepiColombo [on way to Mercury]
2020 – Solar Orbiter [on way to Sun]

KEY

- ○ STILL ACTIVE [AS OF JAN 2024]
- 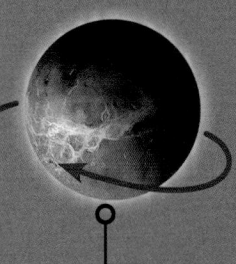 CRASHED
- ✸ FAILED

VENUS, FAILED — 19

- ✸ 1961 – Tyazhely Sputnik
- ✸ 1961 – Venera 1
- ✸ 1962 – Mariner 1
- ✸ 1962 – Sputnik 19
- ✸ 1962 – Sputnik 20
- ✸ 1962 – Sputnik 21
- ✸ 1936 – Cosmos 21
- ✸ 1964 – Venera 1964A
- ✸ 1964 – Venera 1964B
- ✸ 1964 – Cosmos 27
- ✸ 1964 – Zond 1
- ✸ 1965 – Cosmos 96
- ✸ 1965 – Venera 1965A
- ✸ 1966 – Venera 2
- ✸ 1966 – Venera 3*
- ✸ 1967 – Kosmos 167
- ✸ 1970 – Cosmos 359
- ✸ 1972 – Cosmos 482
- ✸ 2010 – Shin'en

MARS — 31

- 1965 – Mariner 4
- 1969 – Mariner 6
- 1969 – Mariner 7
- 1971 – Mariner 9 [first craft to orbit another planet]**
- 1971 – Mars 2**
- 1971 – Mars 3**
- 1974 – Mars 5**
- 1974 – Mars 6**
- 1974 – Mars 7
- 1976 – Viking 1**
- 1976 – Viking 2**
- 1989 – Phobos 2
- ○ 1997 – Mars Pathfinder**
- 1997 – Mars Global Surveyor
- ○ 2001 – Mars Odyssey
- ○ 2003 – Mars Express
- ○ 2004 – MER-A "Spirit"**
- ○ 2004 – MER-B "Opportunity"**
- ○ 2006 – Mars Reconnaissance Orbiter
- 2007 – Rosetta
- 2008 – Phoenix**
- 2009 – Dawn
- ○ 2011 – MSL Curiosity**
- 2014 – Mangalyaan / Mars Orbiter Mission
- ○ 2014 – MAVEN
- ○ 2016 – ExoMars Trace Gas Orbiter
- 2018 – InSight**
- 2020 – Emirates Mars Mission
- ○ 2020 – Tianwen-1**
- ○ 2020 – Mars 2020**
- ○ 2023 – Psyche [on way to asteroid 16 Psyche]

**MARTIAN PROBES

These probes landed or crashed on the surface. See **MARS > MISSIONS TO MARS** | 94 |

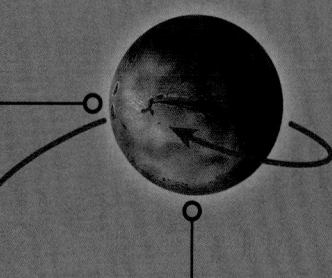

2. MERCURY

MARS, FAILED — 19

- ✸ 1960 – Mars 1M No. 1
- ✸ 1960 – Mars 1M No. 2
- ✸ 1962 – Mars 1962A
- ✸ 1962 – Mars 1962B
- ✸ 1963 – Mars 1
- ✸ 1964 – Mariner 3
- ✸ 1965 – Zond 2
- ✸ 1969 – Mars 1969A
- ✸ 1969 – Mars 1969B
- ✸ 1971 – Mariner 8
- ✸ 1971 – Kosmos 419
- ✸ 1974 – Mars 4
- ✸ 1988 – Phobos 1
- ✸ 1992 – Mars Observer
- ✸ 1996 – Mars 96b
- ✸ 1999 – Mars Climate Orbiter
- ✸ 1999 – Mars Polar Lander**
- ✸ 2003 – Nozomi
- ✸ 2011 – Fobos-Grunt

MARS ROVERS

These probes successfully landed planetary rovers on the surface. See **MARS > SOJOURNER, SPIRIT, CURIOSITY ETC.** | 96-101 |

Not to scale

03.VENUS

MANTLE

* Rocky, 2,840 km thick

ATMOSPHERE

* Mostly carbon dioxide

CORE

* Liquid metallic outer core but, unlike Earth, Venus does not generate a magnetic field because it does not spin fast enough
* Diameter 6,000 km

LIQUID METALLIC OUTER CORE

CRUST = 70 KM

CRUST

* Few impact craters, suggesting "recent" surface (about 500 million years-old) due to massive volcanism

VOLCANOES

* More than 1,700 large, central-vent volcanoes

SURFACE PRESSURE

* 90x Earth

CLOUD

* Reaching 60–70 km above surface, high, thick clouds of concentrated sulphuric acid, formed from sulphur dioxide and water vapour produced by volcanoes

KEY FACTS

Diameter: 12,100 km
Average distance from Sun: 108 million km
Average distance from Earth: 41 million km
Day: 116 Earth days, 18 hours
Year: 225 Earth days
Moons: 0

TWIN SISTER

Venus is roughly the same size and mass as Earth, with a similar internal structure. But that's where the similarities end. See **EARTH'S TWIN?** 34

EARTH

VENUS

EARTH'S SIZE

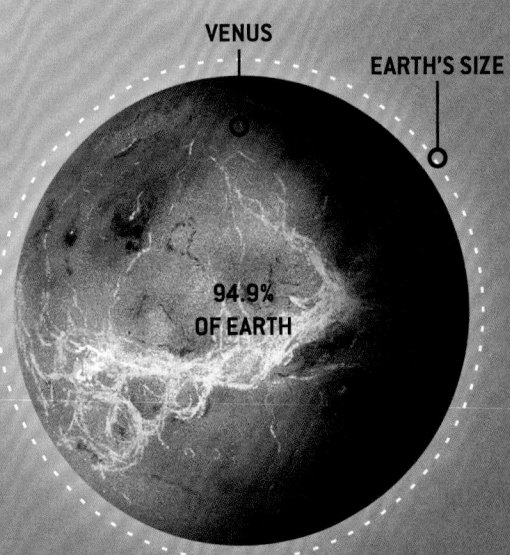

94.9%
OF EARTH

HOT HOT HOT

Carbon dioxide in the atmosphere of Venus has a "greenhouse effect", raising average surface temperature to 462°C — making Venus the hottest planet in the Solar System.

462°C

PLANET OF VOLCANOES

We have not observed any active volcanoes on Venus.

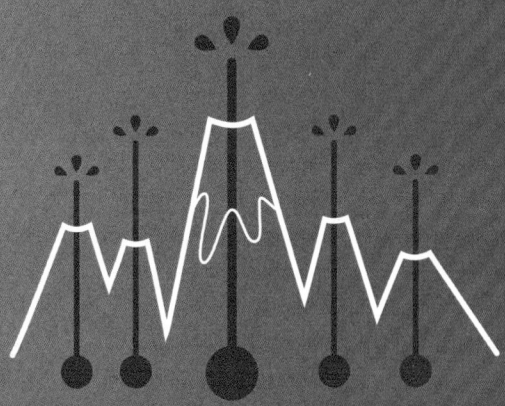

However, changes in the amount of sulphur dioxide in the upper atmosphere, the detection of infrared 'flashes' and the distinctive weathering of some Venusian rocks may all be evidence of some active volcanism.

BIG MAAT

Maat Mons is the biggest volcano on Venus. It is 8 km high and has a diameter of 400 km!

EARTH'S TWIN?

Venus and Earth are similar in size, mass, structure and distance from the Sun – but there the resemblance ends.

WATER

No known liquid water

TEMPERATURE

Constant surface temperature of 462°C

LIFE

No known life

TECTONIC PLATES

No tectonic plates (crust may be too hot and soft)

GEOLOGY

Mostly smooth, volcanic plains within 1 km of median height, unchanged for 300–600 million years

VOLCANOES

The most volcanic planet in the Solar System with more than 1,700 major volcanoes and between 100,000 and 1,000,000 minor ones, none are known to be active — but some might be

MAGNETIC FIELD

Weak magnetic field offers little protection from cosmic radiation

AIR PRESSURE

Air pressure at surface: a crushing 9,200 kPa

OBLIQUITY

Minimal obliquity (axial tilt) so no seasons

CLOUD

Thick cloud of sulphuric acid

The thick cloud on Venus lets in some light. From the surface, the sky is tangerine-colour and there's enough light to see shadows.

WATER

70% of surface covered in liquid water

TEMPERATURE

Average 14°C, but temperatures range from −89.2°C to 70.7°C

LIFE

Abundance of life

TECTONIC PLATES

Tectonic plates which aid heat loss

GEOLOGY

Varied geological features and continual reshaping through tectonics and erosion

VOLCANOES

~1,500 active volcanoes

MAGNETIC FIELD

Liquid outer core generates protective magnetic fields

OBLIQUITY

Obliquity = 23.4° so periodic seasons

AIR PRESSURE

Air pressure at surface: 101.3 kPa

CLOUD

Intermittent clouds of water vapour

KEEP YOUR DISTANCE

Earth is in the habitable or "Goldilocks" zone, neither too hot nor too cold for liquid water, which we think may be vital for life!

EARTH:

1 AU

MERCURY

(Distance from sun):

0.39 AU

SUN

0.38 AU

0.725 AU

0.95 AU

1.24 AU

1.37 AU

VENUS:

0.72 AU

MARS:

1.5 AU

 Estimates of where goldilocks zones may lie are based on factors such as the thickness and composition of a planet's atmosphere, which can affect whether conditions on its surface are right for water.

GOLDILOCKS ZONE

Most commonly used estimate of habitable zone, by James F Kasting, Daniel P Whitmire and Ray T Reynolds (1993)

 Stars cooler than the Sun will have closer goldilocks zones and hotter stars will have zones further out.

FIRST ESTIMATE

First estimate of goldilocks zone, by Stephen H Dole (1964)

EXTREME EDGE

Estimate by Raymond Pierrehumbert and Eric Gaidos (2011) that, in certain conditions, the zone could extend beyond Saturn

TO 10 AU

3.0 AU

INNER EDGE

Estimate by Andras Zsom, Sara Seager and Julien De Wit (2013)

OUTER EDGE

Estimate by MJ Fogg (1992)

Understanding the reach of the Goldilocks zone helps us estimate the chances of finding life on 'exoplanets' – planets orbiting other stars!

EXPLORERS

All the probes we've landed or crashed on our twin planet.

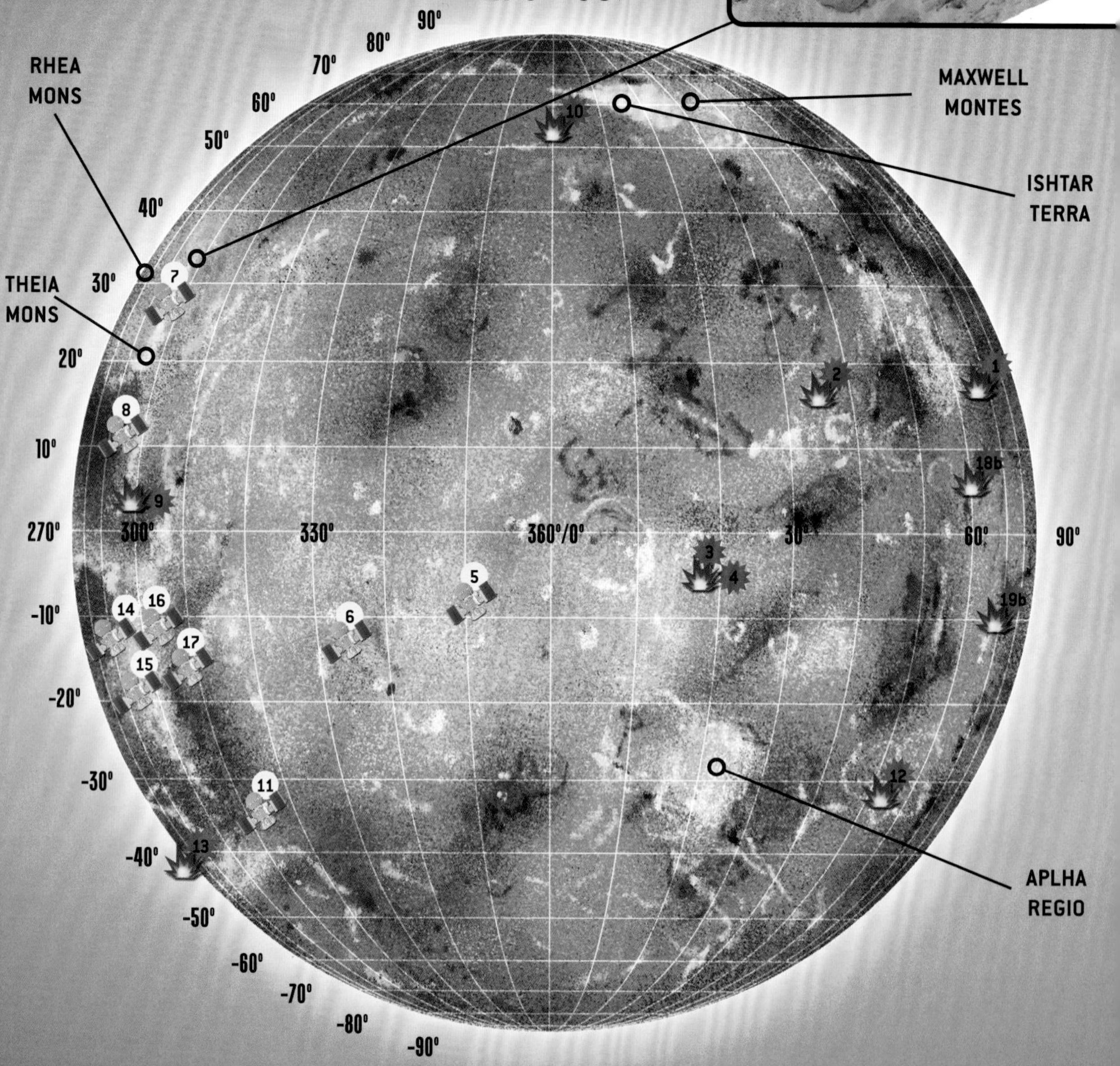

270°–90°

RHEA MONS

THEIA MONS

MAXWELL MONTES

ISHTAR TERRA

APLHA REGIO

KEY TO LANDINGS

Crash

Land

1. Venera 3: 1 March 1966

2. Venera 4: 18 October 1967 — remained sending data down to altitude of 25 km

3. Venera 5: 16 May 1969 — sent data down to altitude of 26 km

4. Venera 6: 17 May 1969 — sent data down to altitude of 10-12 km

5. Venera 7: 15 December 1970 — sent data from surface for 23 min

6. Venera 8: 22 July 1972 — sent data from surface for 50 min

7. Venera 9: 22 October 1975 — sent data including first images from surface for 53 min

8. Venera 10: 25 October 1975 — sent data from surface for 65 min

9. Pioneer Venus, Large probe: 9 December 1978 — sent data until crash

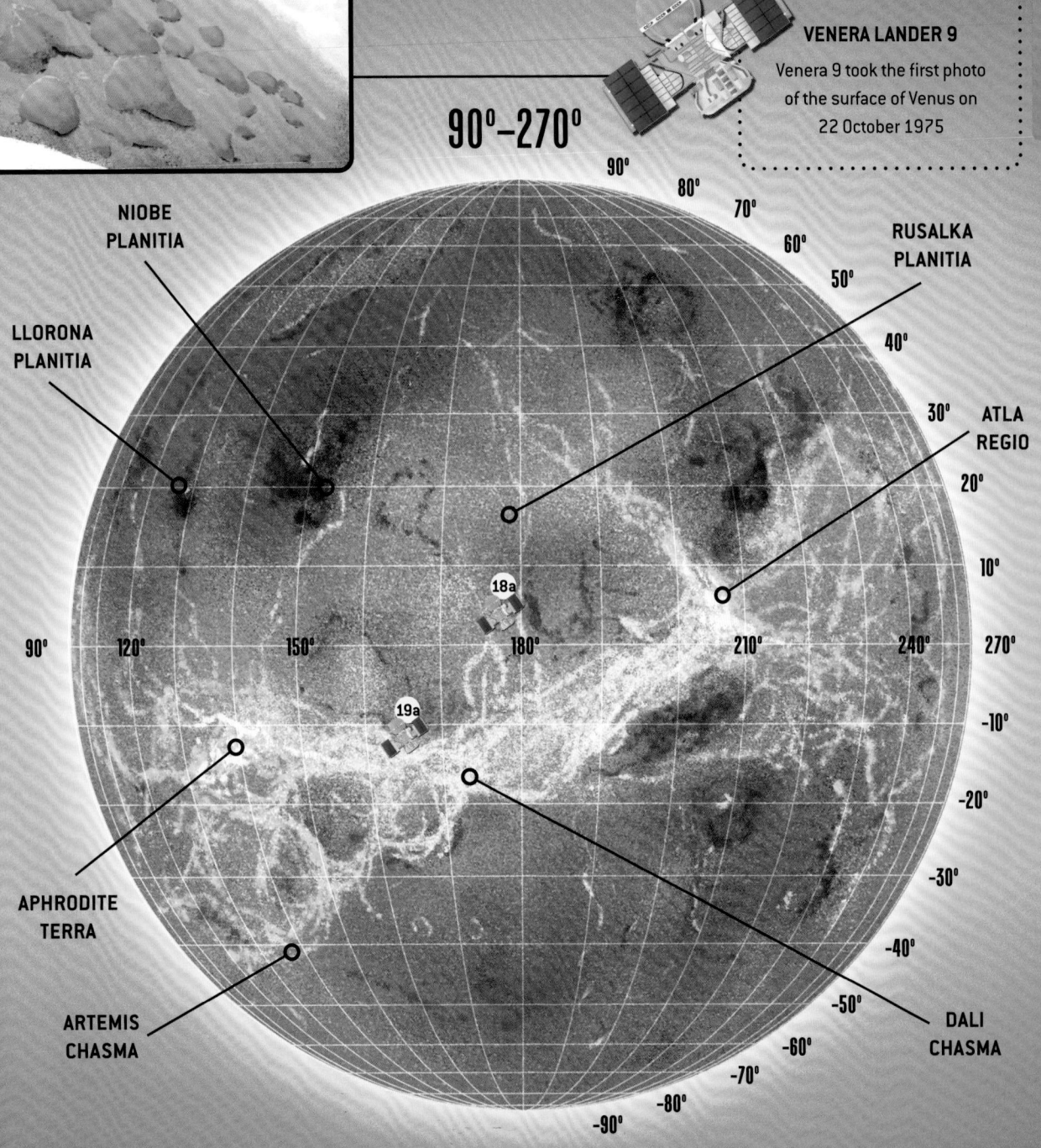

VENERA LANDER 9

Venera 9 took the first photo of the surface of Venus on 22 October 1975

90°–270°

90°
80°
70°
60°
50°
40°
30°
20°
10°
270°
240°
210°
180°
150°
120°
90°
-10°
-20°
-30°
-40°
-50°
-60°
-70°
-80°
-90°

NIOBE PLANITIA

LLORONA PLANITIA

RUSALKA PLANITIA

ATLA REGIO

18a

19a

APHRODITE TERRA

ARTEMIS CHASMA

DALI CHASMA

10. Pioneer Venus, North probe: 9 December 1978 — sent data until crash

11. Pioneer Venus, Day probe: 9 December 1978 — sent data from surface for 67 min

12. Pioneer Venus, Night probe: 9 December 1978 — sent data until crash

13. Pioneer Venus, Bus: 9 December 1978, — sent data down to altitude of 110 km

14. Venera 12: 21 December 1978, sent data from surface for 110 min

15. Venera 11: 25 December 1978 — sent data from surface for 95 min

16. Venera 13: 1 March 1982 — sent data from surface for 127 min

17. Venera 14: 5 March 1982 — sent data from surface for 57 min

18a. Vega 1, Descent lander: 11 June 1985 — sent data from surface for 56 min

18b. Vega 1, Balloon: 13 June 1985 — last known co-ordinates at altitude of 54 km

19a. Vega 2, Descent lander: 15 June 1985 — sent data from surface for 56 min

19b. Vega 2, Balloon: 17 June 1985 — last known co-ordinates at altitude of 54 km

04.EARTH

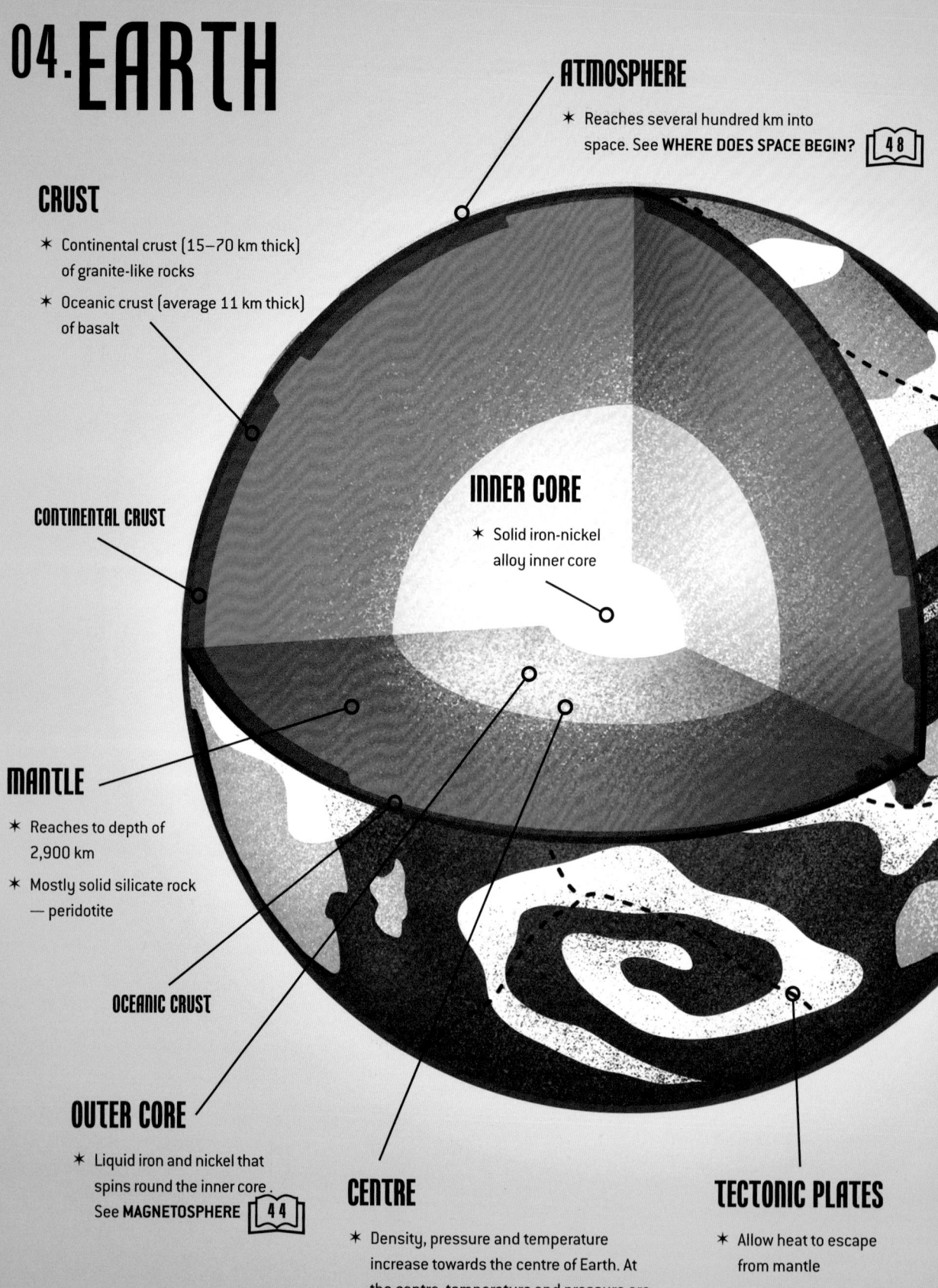

ATMOSPHERE

* Reaches several hundred km into space. See **WHERE DOES SPACE BEGIN?** [4 8]

CRUST

* Continental crust (15–70 km thick) of granite-like rocks
* Oceanic crust (average 11 km thick) of basalt

CONTINENTAL CRUST

INNER CORE

* Solid iron-nickel alloy inner core

MANTLE

* Reaches to depth of 2,900 km
* Mostly solid silicate rock — peridotite

OCEANIC CRUST

OUTER CORE

* Liquid iron and nickel that spins round the inner core . See **MAGNETOSPHERE** [4 4]

CENTRE

* Density, pressure and temperature increase towards the centre of Earth. At the centre, temperature and pressure are similar to the surface of the Sun!

TECTONIC PLATES

* Allow heat to escape from mantle

KEY FACTS

Diameter: 12,750 km
Average distance from Sun: 150 million km
Day: 1 Earth day (24 hours)
Year: 365.25 Earth days
Moons: 1

LIFE ON EARTH

Earth is the only place in the universe where we know life exists. And what a *lot* of life …

= 1 BILLION

= 1,000 TRILLION

= 1,000 TRILLION

**7.7 BILLION
HUMANS**

**10,000 TRILLION
ANTS**

**100,000 TRILLION
INSECTS**

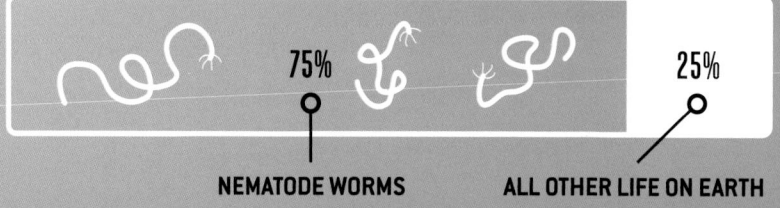

75%

25%

NEMATODE WORMS

ALL OTHER LIFE ON EARTH

COMPOSITION
OF EARTH

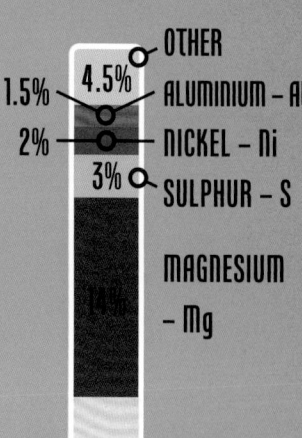

1.5% **4.5%** — OTHER
— ALUMINIUM – Al
2% — NICKEL – Ni
3% — SULPHUR – S

14% MAGNESIUM
– Mg

15% SILICON – Si

30% OXYGEN – O$_2$

30% IRON – Fe

41

LIVING PLANET

So far, life has only been found on Earth. We think life exists here due to a combination of factors.

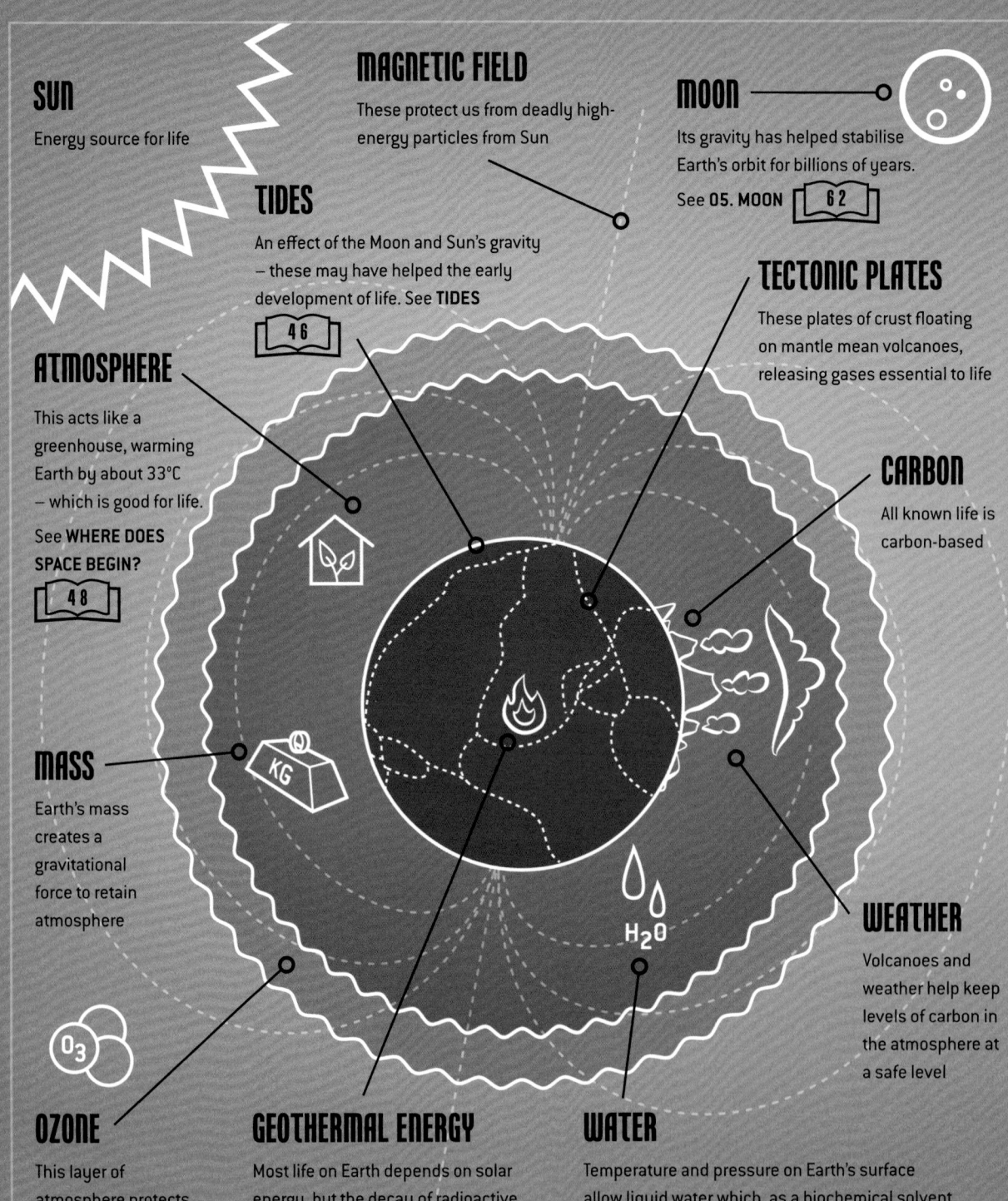

SUN

Energy source for life

MAGNETIC FIELD

These protect us from deadly high-energy particles from Sun

MOON

Its gravity has helped stabilise Earth's orbit for billions of years.

See **05. MOON** `62`

TIDES

An effect of the Moon and Sun's gravity – these may have helped the early development of life. See **TIDES** `46`

TECTONIC PLATES

These plates of crust floating on mantle mean volcanoes, releasing gases essential to life

ATMOSPHERE

This acts like a greenhouse, warming Earth by about 33°C – which is good for life.

See **WHERE DOES SPACE BEGIN?** `48`

CARBON

All known life is carbon-based

MASS

Earth's mass creates a gravitational force to retain atmosphere

WEATHER

Volcanoes and weather help keep levels of carbon in the atmosphere at a safe level

OZONE

This layer of atmosphere protects us from the Sun's ultraviolet radiation

GEOTHERMAL ENERGY

Most life on Earth depends on solar energy, but the decay of radioactive isotopes in Earth's crust gives some life an alternative energy source

WATER

Temperature and pressure on Earth's surface allow liquid water which, as a biochemical solvent, seems essential to all life. 70% of Earth's surface is submerged in water!

LIFE ON EARTH

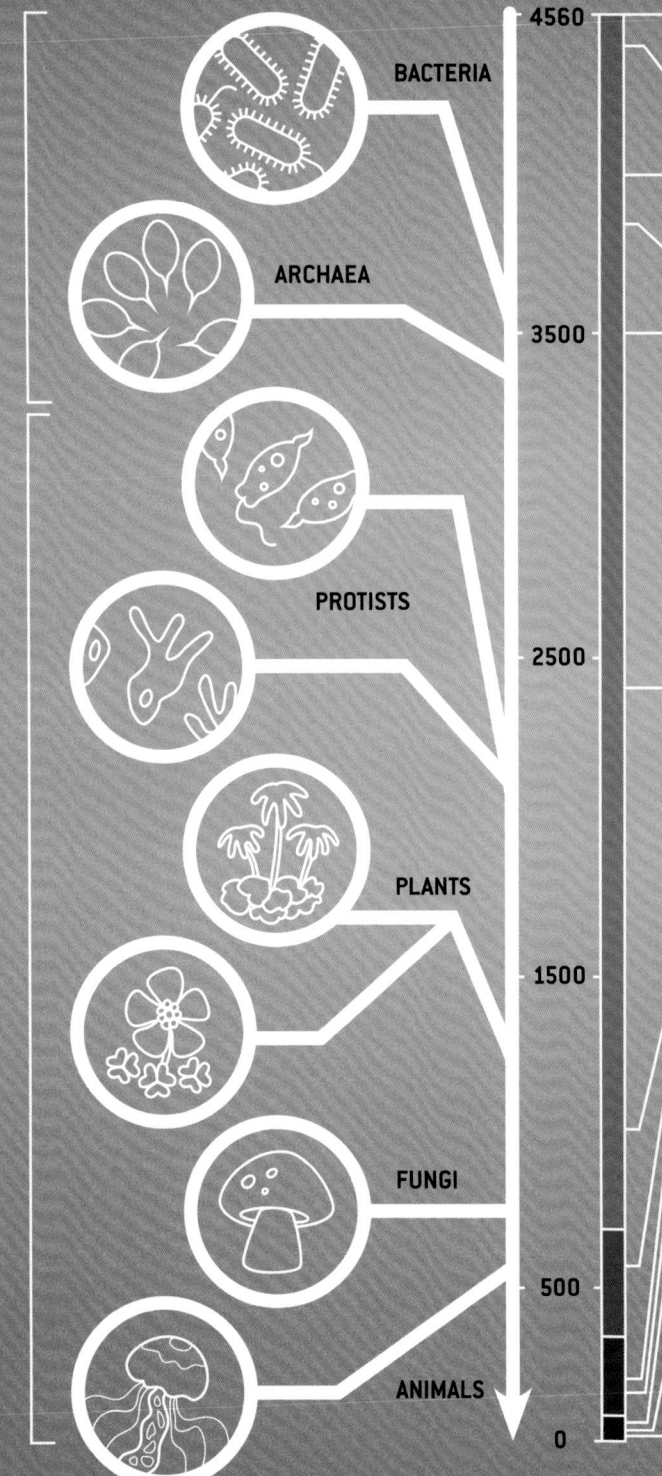

PROKARYOTES

EUKARYOTES

BACTERIA

ARCHAEA

PROTISTS

PLANTS

FUNGI

ANIMALS

MILLIONS OF YEARS AGO

4560

3500

2500

1500

500

0

PRECAMBRIAN

PALEOZOIC **MESOZOIC** **CENZOIC**

4, 560, 000, 000 — Earth forms

4, 400, 000, 000 — Crust solidifies as
Earth cools; evidence of liquid water

4, 000, 000, 000 — Estimated first life on Earth

3, 850, 000, 000 — First evidence of life

3, 500, 000, 000 — First fossilised
evidence of bacteria

2, 400, 000, 000 — First Eukaryotic life
with complex cell systems

1, 000, 000, 000 — First multicellular animals

545, 000, 000 — "Cambrian explosion",
evolution of modern animal bodies

200, 000, 000 — First mammals

140, 000, 000 — First flowers

66, 000, 000 — Extinction of dinosaurs
and ¾ of all animal and plant species

300, 000 — First fossils of *Homo sapiens*,
in what's now Ethiopia

150, 000 — First modern humans left Africa

MAGNETOSPHERE

Earth's magnetic field is generated by electric currents in the planet's liquid outer core. It protects us from solar wind and cosmic rays.

SOLAR WIND

Stream of charged particles, moving at 200 to 1,000 km per second, which could damage Earth's atmosphere.

PLASMASPHERE

Doughnut-shaped region of low-energy particles rotating with Earth. Partially overlaps with Van Allen belts.

SOLAR WIND

MAGNETIC FIELD LINES

VAN ALLEN INNER BELT

VAN ALLEN OUTER BELT

VAN ALLEN BELTS

Two tyre-shaped belts of high-energy particles from solar wind and cosmic rays, trapped by the magnetic fields and so prevented from damaging Earth's atmosphere.

BOW SHOCK

Solar wind slowed by Earth's magnetic field and mostly deflected around planet.

COSMIC RAYS

High energy charged particles — mostly originating outside the Solar System — are also deflected.

Not to scale

MAGNETOPAUSE

Boundary where pressure of solar wind and of magnetic field balance. The amount of solar wind varies, so this boundary does too.

MAGNETOSPHERE

Magnetic field generated, mostly, by the turning of the Earth's liquid iron outer core. In higher latitudes, the shape of the magnetosphere is distorted by solar wind.

AURORAE

Some charged particles reach the magnetosphere and, where they collide with atoms in the ionosphere at the poles, can create colourful lights in the sky, the aurorae.

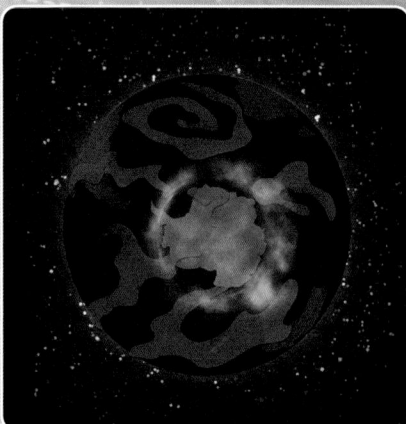

ATMOSPHERE

Relatively thin, fragile layer of gases vital to life on Earth.

MAGNETOTAIL

Twin "tails" of magnetosphere extending 100x diameter of Earth.

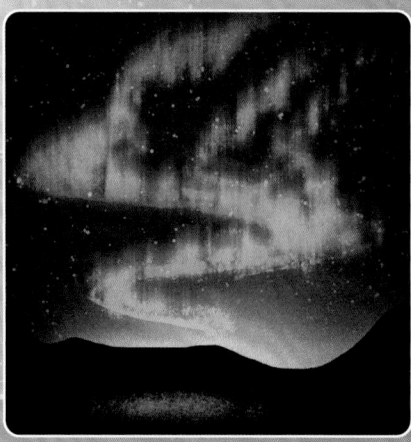

MAGNETOSHEATH

Area of "shocked" solar wind between bow shock and magnetopause.

TIDES

71% of the Earth's surface is covered by water, which is affected by the Moon and Sun.

GRAVITY

All things with mass or energy *gravitate* or are attracted toward one another. The effects are especially noticeable if at least one of the objects is massive.

HOW THE MOON CREATES TIDES

The gravitational forces exerted by the Moon make the seas on Earth rise and fall in regular sequence.

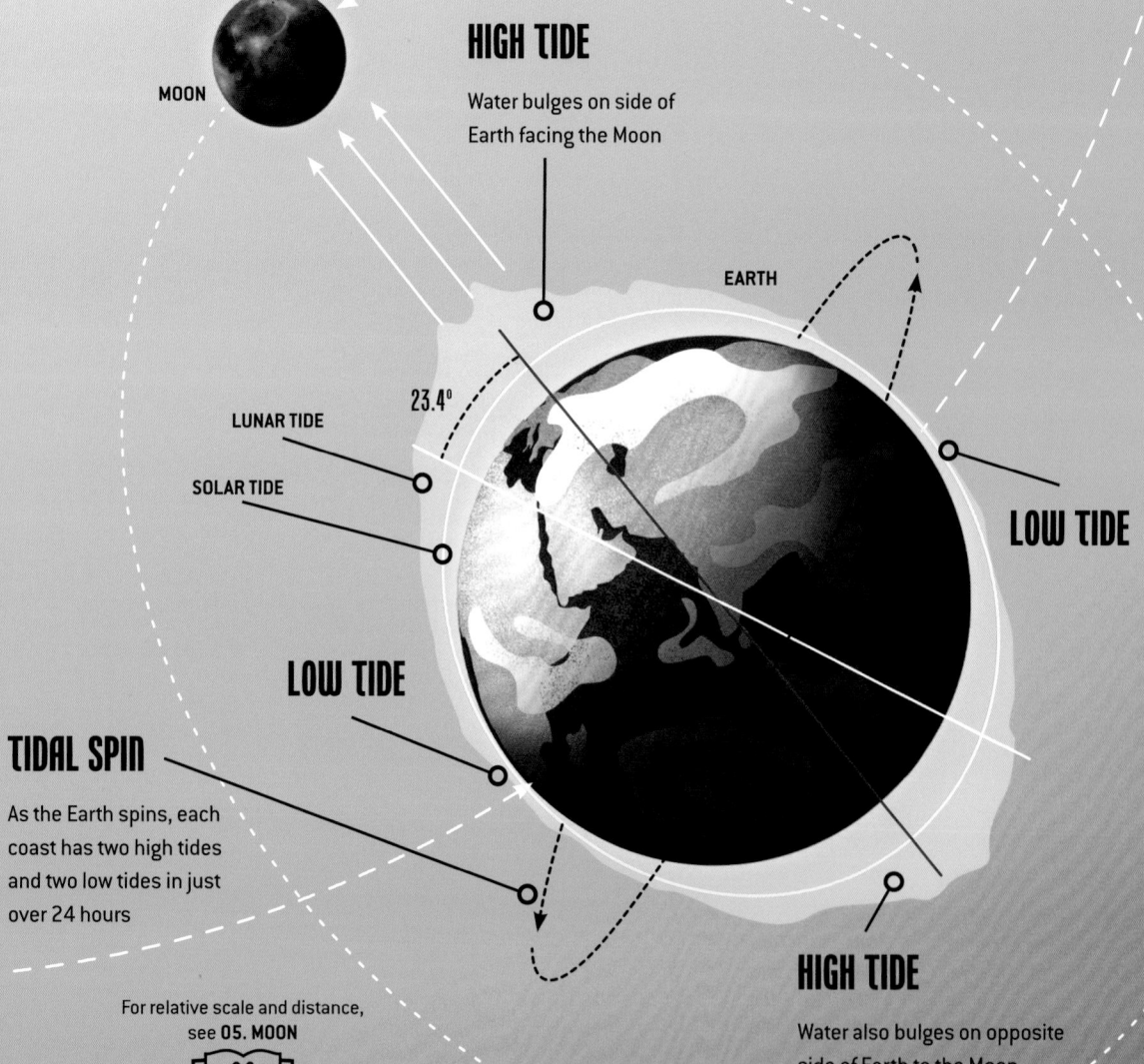

MOON

HIGH TIDE

Water bulges on side of Earth facing the Moon

EARTH

23.4°

LUNAR TIDE

SOLAR TIDE

LOW TIDE

LOW TIDE

TIDAL SPIN

As the Earth spins, each coast has two high tides and two low tides in just over 24 hours

HIGH TIDE

Water also bulges on opposite side of Earth to the Moon

For relative scale and distance, see 05. MOON

62

HOW THE SUN AFFECTS TIDES

The Sun affects the sea with about half of the effect of the Moon.

SPRING TIDES

When the Sun and Moon are aligned at "full Moon" or "new Moon" (every 14 days), their gravitational influence works together to create higher high tides and lower low tides.

1. New Moon

2. Full Moon

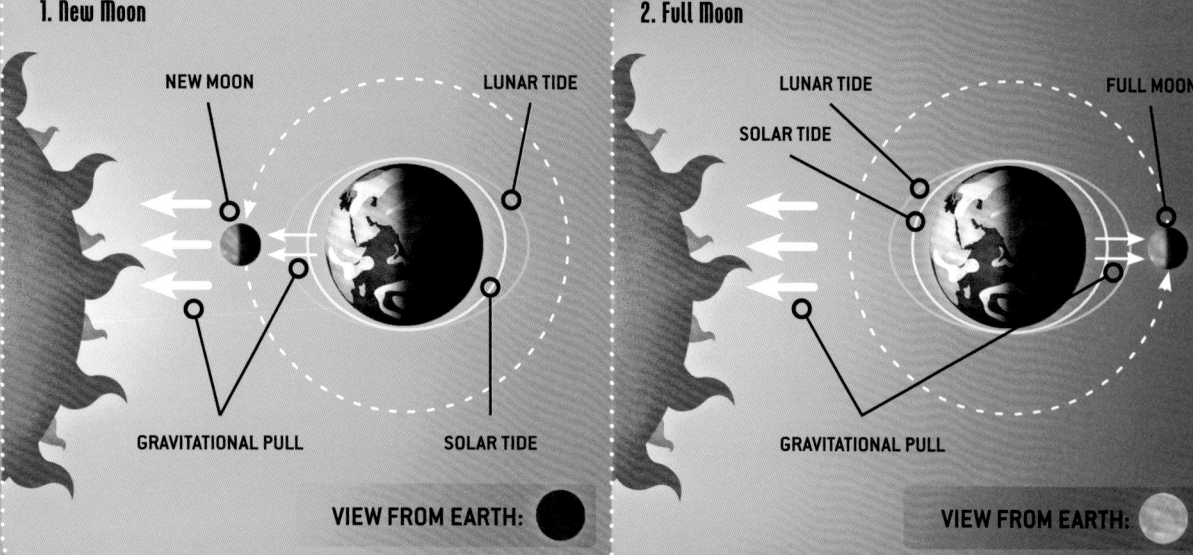

NEW MOON

LUNAR TIDE

GRAVITATIONAL PULL

SOLAR TIDE

VIEW FROM EARTH:

LUNAR TIDE

SOLAR TIDE

FULL MOON

GRAVITATIONAL PULL

VIEW FROM EARTH:

NEAP TIDES

When the Sun is at 90° to the Moon (at "First Quarter" and "Last Quarter"), their gravitational forces work against each other: higher low tides and lower high tides. See **MOON > PHASES** 66

3. First Quarter

4. Last Quarter

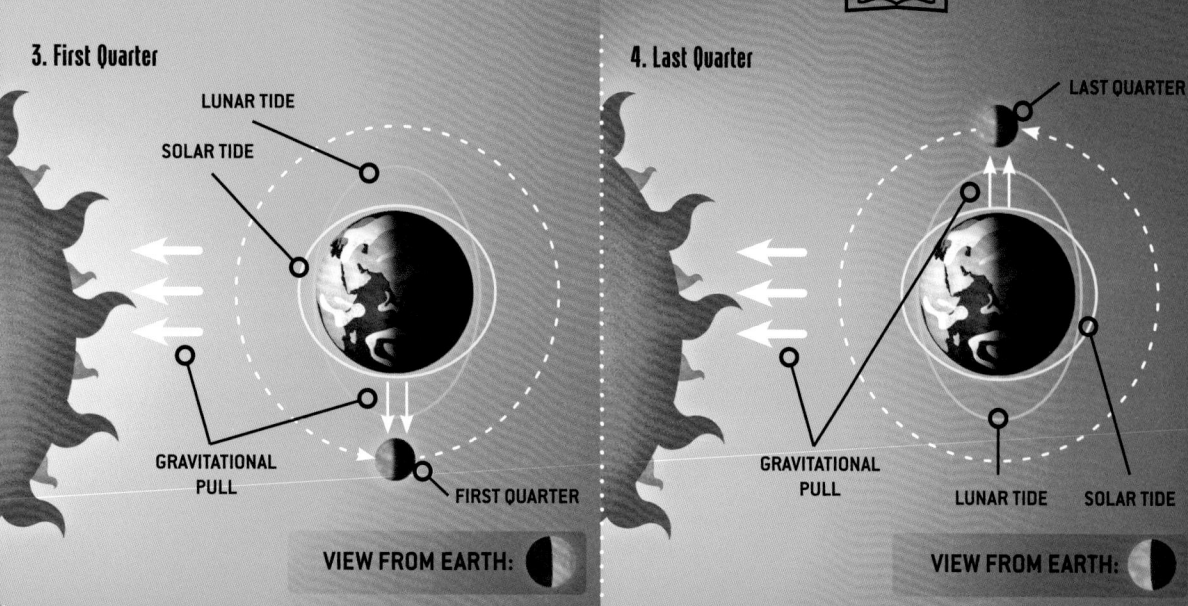

LUNAR TIDE

SOLAR TIDE

GRAVITATIONAL PULL

FIRST QUARTER

VIEW FROM EARTH:

LAST QUARTER

GRAVITATIONAL PULL

LUNAR TIDE

SOLAR TIDE

VIEW FROM EARTH:

WHERE DOES SPACE BEGIN?

*The further from Earth's surface, the less dense the atmosphere —
but there's no clear boundary between us and space.*

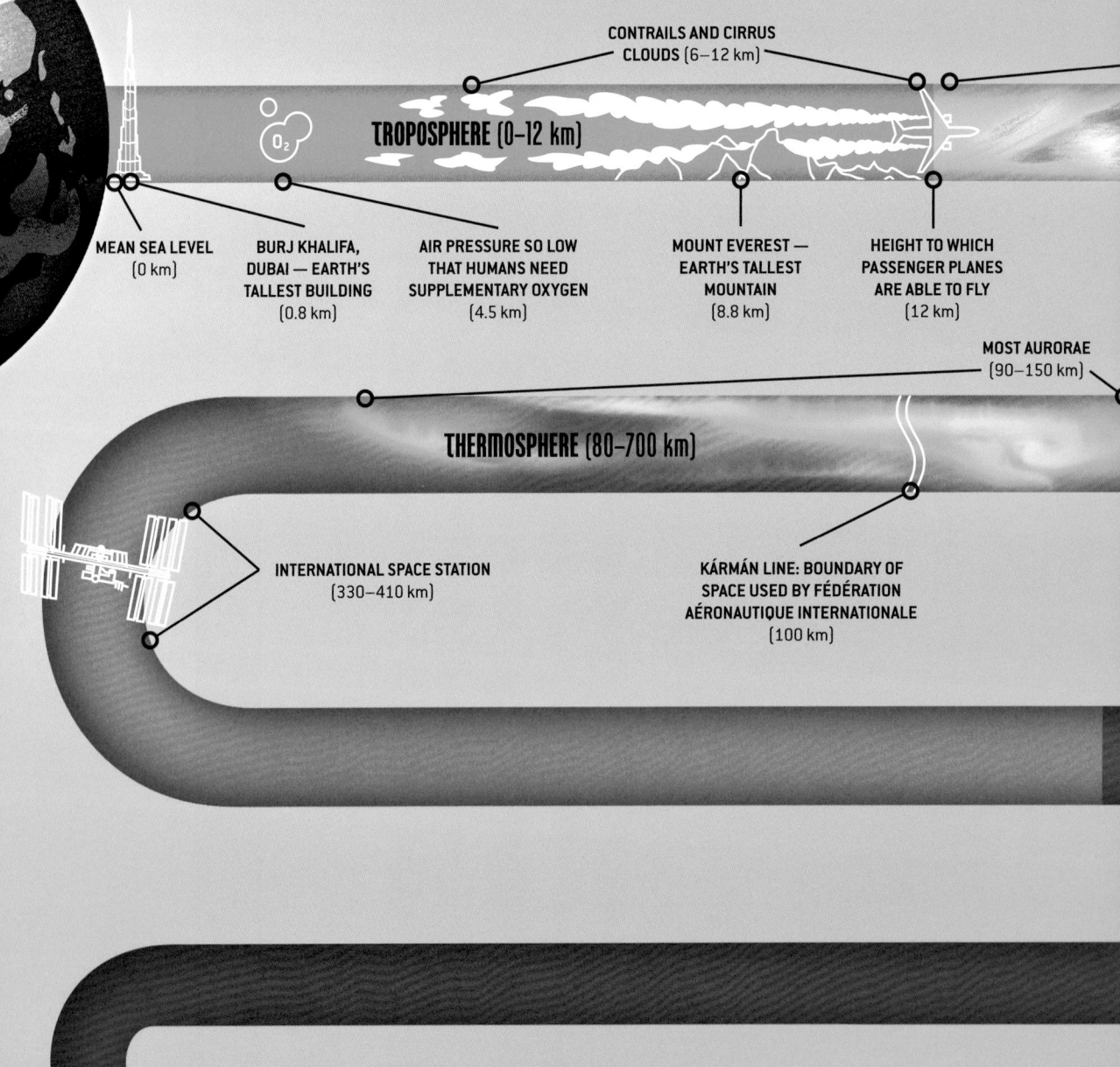

CONTRAILS AND CIRRUS
CLOUDS (6–12 km)

O₂

TROPOSPHERE (0–12 km)

MEAN SEA LEVEL
(0 km)

BURJ KHALIFA,
DUBAI — EARTH'S
TALLEST BUILDING
(0.8 km)

AIR PRESSURE SO LOW
THAT HUMANS NEED
SUPPLEMENTARY OXYGEN
(4.5 km)

MOUNT EVEREST —
EARTH'S TALLEST
MOUNTAIN
(8.8 km)

HEIGHT TO WHICH
PASSENGER PLANES
ARE ABLE TO FLY
(12 km)

MOST AURORAE
(90–150 km)

THERMOSPHERE (80–700 km)

INTERNATIONAL SPACE STATION
(330–410 km)

KÁRMÁN LINE: BOUNDARY OF
SPACE USED BY FÉDÉRATION
AÉRONAUTIQUE INTERNATIONALE
(100 km)

GENERALLY AGREED BOUNDARY BETWEEN
EARTH'S ATMOSPHERE AND SPACE
(10,000 km)

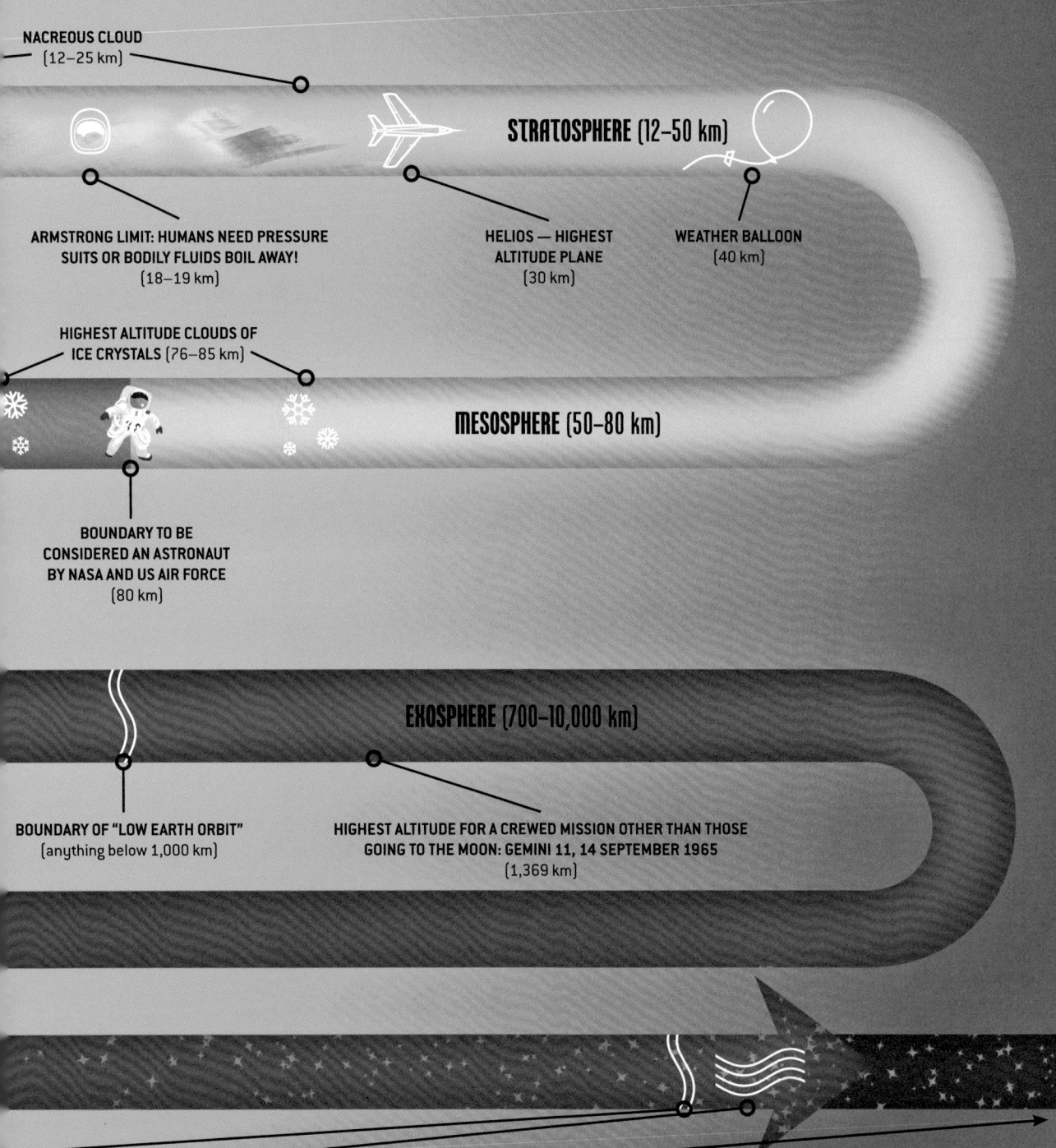

NACREOUS CLOUD
(12–25 km)

STRATOSPHERE (12–50 km)

ARMSTRONG LIMIT: HUMANS NEED PRESSURE
SUITS OR BODILY FLUIDS BOIL AWAY!
(18–19 km)

HELIOS — HIGHEST
ALTITUDE PLANE
(30 km)

WEATHER BALLOON
(40 km)

HIGHEST ALTITUDE CLOUDS OF
ICE CRYSTALS (76–85 km)

MESOSPHERE (50–80 km)

BOUNDARY TO BE
CONSIDERED AN ASTRONAUT
BY NASA AND US AIR FORCE
(80 km)

EXOSPHERE (700–10,000 km)

BOUNDARY OF "LOW EARTH ORBIT"
(anything below 1,000 km)

HIGHEST ALTITUDE FOR A CREWED MISSION OTHER THAN THOSE
GOING TO THE MOON: GEMINI 11, 14 SEPTEMBER 1965
(1,369 km)

SOLAR WIND
(10,000 km and beyond)

ATMOSPHERIC PARTICLES DETECTED STILL MORE
AFFECTED BY EARTH'S GRAVITY THAN BY SOLAR WIND
(100,000 km)

ESTIMATE OF WHERE EARTH'S GRAVITY NO LONGER AFFECTS
ATMOSPHERIC PARTICLES MORE THAN SOLAR WIND
(190,000 km)

METEOROIDS, METEORS, METEORITES

Rocks and dust that blaze brightly through Earth's atmosphere.

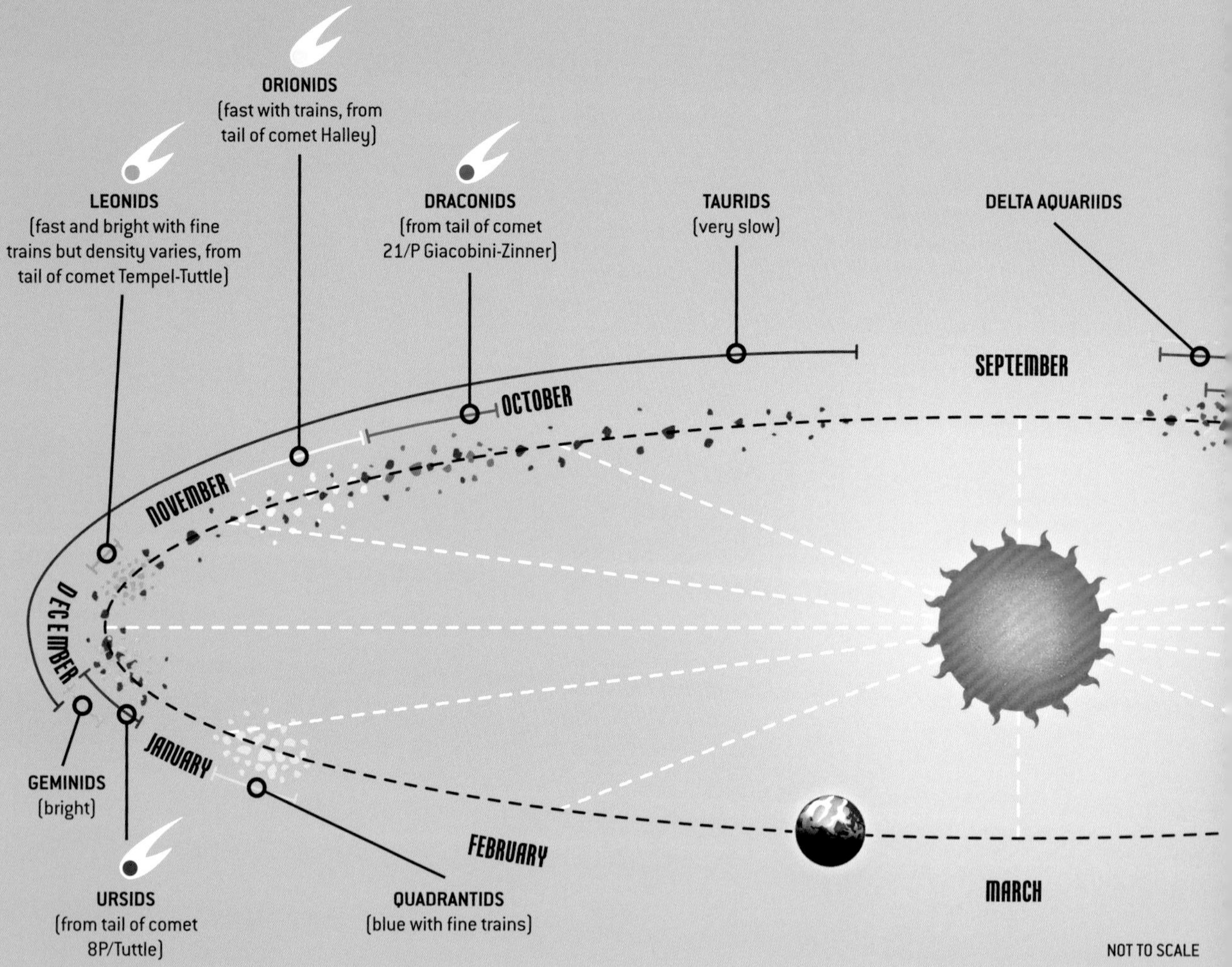

ORIONIDS
(fast with trains, from tail of comet Halley)

LEONIDS
(fast and bright with fine trains but density varies, from tail of comet Tempel-Tuttle)

DRACONIDS
(from tail of comet 21/P Giacobini-Zinner)

TAURIDS
(very slow)

DELTA AQUARIIDS

SEPTEMBER

OCTOBER

NOVEMBER

DECEMBER

JANUARY

FEBRUARY

MARCH

GEMINIDS
(bright)

URSIDS
(from tail of comet 8P/Tuttle)

QUADRANTIDS
(blue with fine trains)

NOT TO SCALE

METEOROIDS...

...are dust and rocks in space and smaller than asteroids. Most are cometary dust but large meteoroids are often bits of asteroid. See **MARS >**
ASTEROID BELT [106]

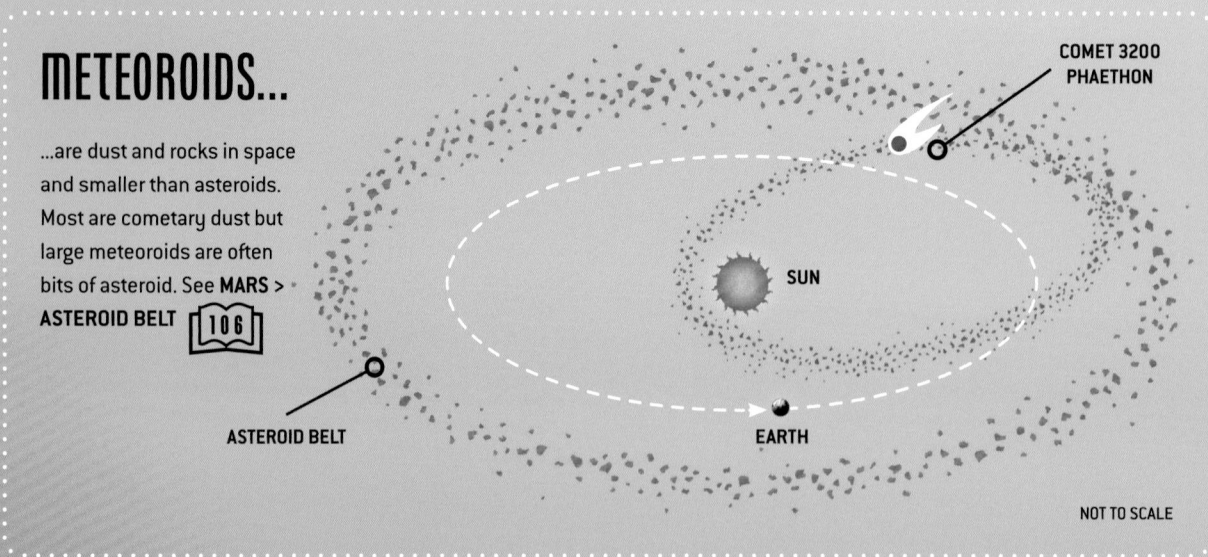

COMET 3200 PHAETHON

SUN

ASTEROID BELT

EARTH

NOT TO SCALE

METEOR SHOWERS

Random, "sporadic meteors" can be seen about 7 times an hour in a dark sky.

But there are also regular, dense "meteor showers" through the year as Earth's orbital path crosses the dust particles left behind by comets.

Meteors in showers appear to radiate from a specific point or "radiant" in the sky, and are named after the constellation of stars in which the radiant is seen — for example the "Orionids" in the constellation of Orion.

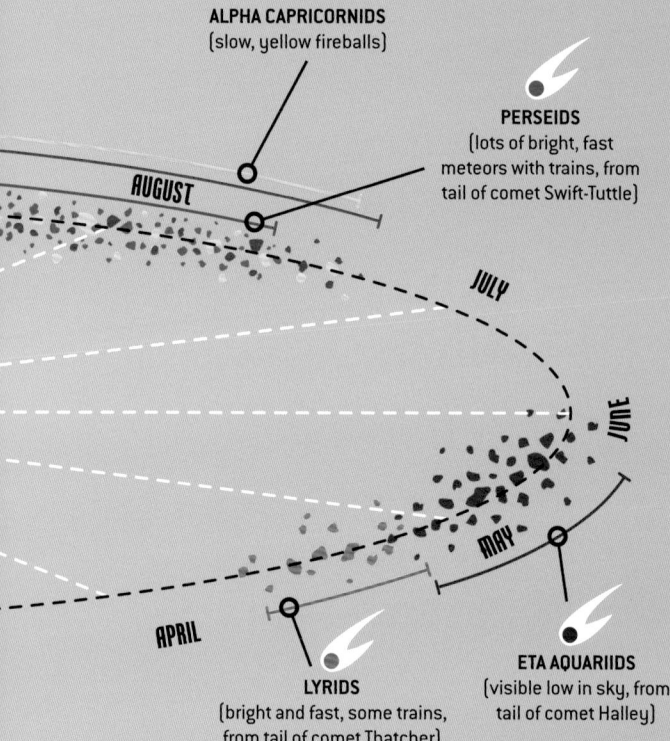

ALPHA CAPRICORNIDS
(slow, yellow fireballs)

PERSEIDS
(lots of bright, fast meteors with trains, from tail of comet Swift-Tuttle)

AUGUST

JULY

JUNE

MAY

APRIL

LYRIDS
(bright and fast, some trains, from tail of comet Thatcher)

ETA AQUARIIDS
(visible low in sky, from tail of comet Halley)

METEORS...

...are meteoroids that come in contact with Earth and usually burn up brightly in the atmosphere.

The brightest meteors leave visible trails or "trains" of hot gas behind them.

Some meteors are so bright they can be seen during the day and are known as "fireballs".

MICROMETEORITES...

...originate from space and are less than 0.5 mm in size. They can be found all over the world, including on urban rooftops.

METEORITES...

...are fragments of large meteors that survive to reach the ground.

About 31,000 meteorites have been collected from all over the world's surface:

* 25 originating from the Moon
* 30 from Mars
* 2 from unknown dwarf planet
* Vast majority from asteroids

They are classified as either iron, stony-iron or stony, and most are 4.5 billion years old.

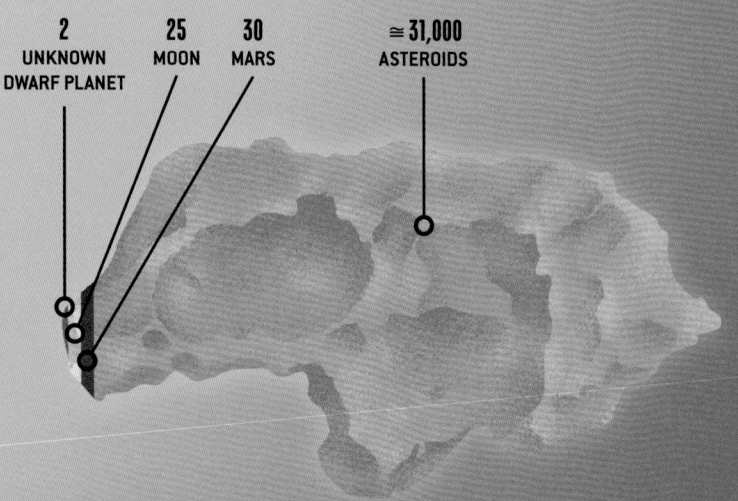

2
UNKNOWN DWARF PLANET

25
MOON

30
MARS

≅ 31,000
ASTEROIDS

POTENTIALLY HAZARDOUS OBJECTS

Asteroids, comets and other objects can come dangerously close to Earth!
These ones are larger than 50 metres in size.

NEAR-EARTH OBJECTS

* 19,470 known near-Earth objects (NEOs), which at closest approach are within 1.3 AU of Sun
* 19,363 (99.5%) of them = asteroids
* 1,955 (10%) of them > 140 m and considered potentially hazardous

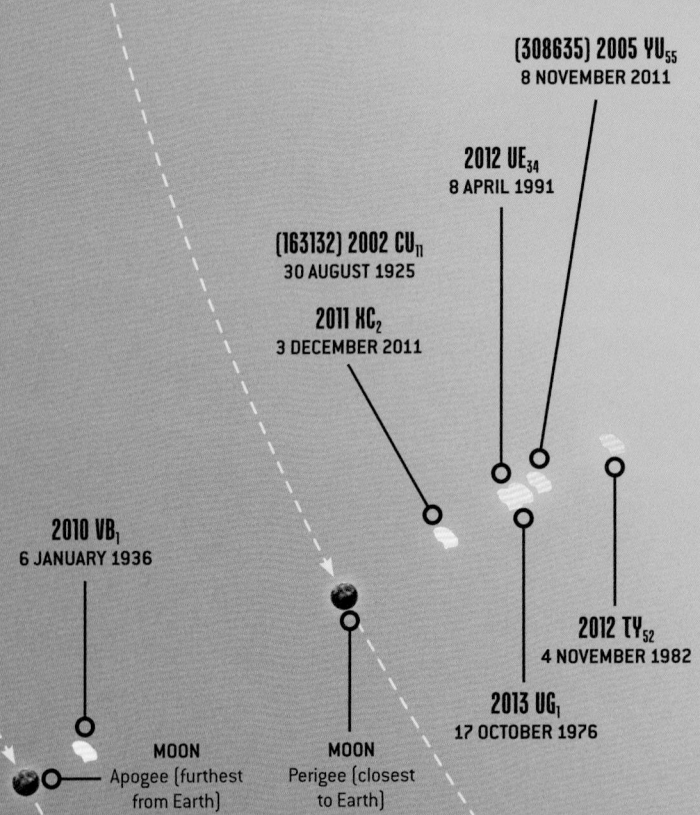

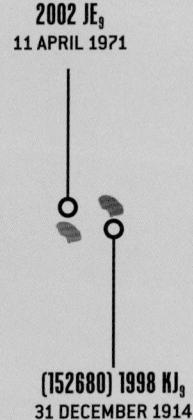

2002 JE$_9$
11 APRIL 1971

(308635) 2005 YU$_{55}$
8 NOVEMBER 2011

2012 UE$_{34}$
8 APRIL 1991

(163132) 2002 CU$_{11}$
30 AUGUST 1925

2011 XC$_2$
3 DECEMBER 2011

(152680) 1998 KJ$_9$
31 DECEMBER 1914

2010 VB$_1$
6 JANUARY 1936

2012 TV$_{52}$
4 NOVEMBER 1982

2013 UG$_1$
17 OCTOBER 1976

MOON
Apogee (furthest from Earth)

MOON
Perigee (closest to Earth)

ASTEROID SIZES NOT TO SCALE

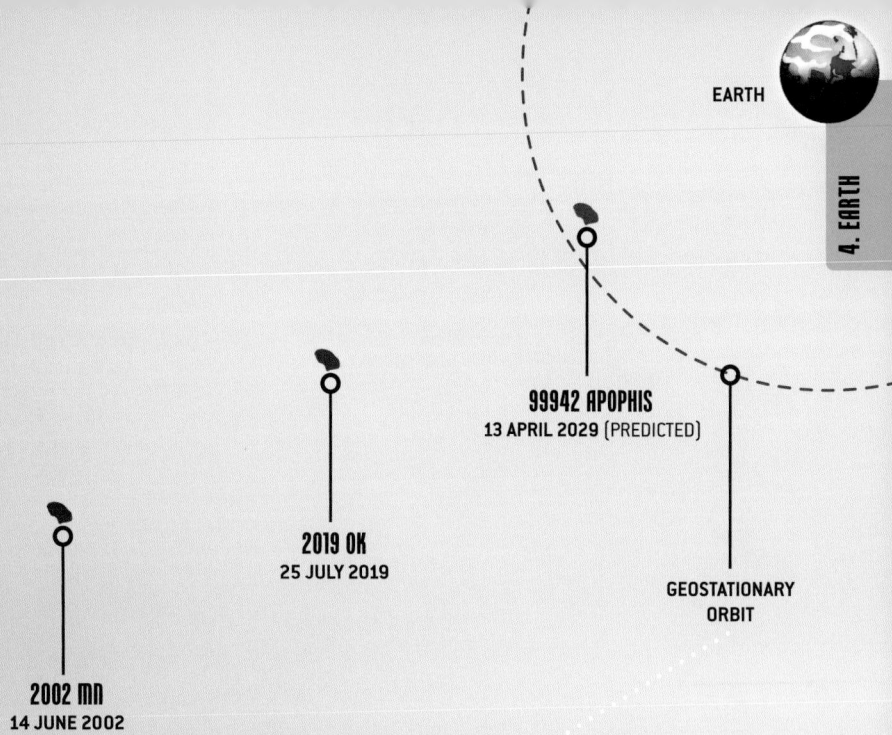

99942 APOPHIS
13 APRIL 2029 (PREDICTED)

2019 OK
25 JULY 2019

GEOSTATIONARY
ORBIT

2002 MN
14 JUNE 2002

GEOSTATIONARY ORBIT

At this distance, an object takes 24 hours to
complete an orbit of the Earth, so effectively
remains in a fixed position above a point on Earth.

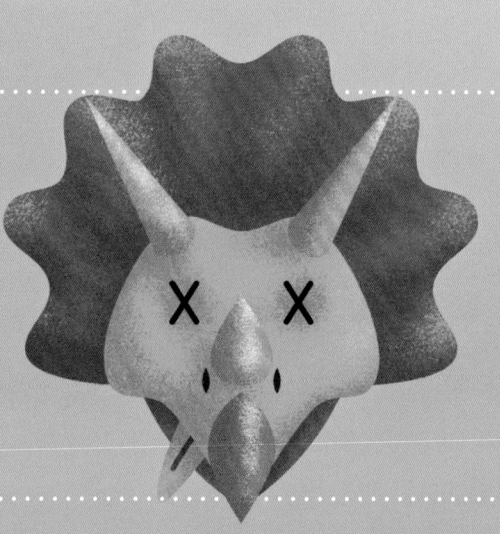

DINO-MITE

Evidence suggests that a
comet or asteroid 10–15 km
wide hit the Earth 66 million
years ago and wiped out
75% of all living species —
including the dinosaurs.

ANIMALS IN SPACE

As well as humans, we've launched a wide — and surprising — range of creatures into space to study the impact on them.

FRUIT FLY
20 FEBRUARY 1947

WASP
1967

FLOUR BEETLE
1967

WINE FLY
1968

MEALWORM
1968

NEMATODE
1972

MUMMICHOG
1975

GARDEN SPIDER
1975

ZEBRA DANIO
1976

RHESUS MONKEY
14 JUNE 1949

SQUIRREL MONKEY
13 DECEMBER 1958

MOUSE
31 AUGUST 1950

DOG
22 JULY 1951

HUMAN
5 MAY 1961

RABBIT
2 JULY 1959

FROG
19 SEPTEMBER 1959

CHIMPANZEE
31 JANUARY 1961

CAT
1963

GUINEA PIG
1961

RAT
19 AUGUST 1960

NEWT
1985

BRINE SHRIMP
1990

JELLYFISH
1991

ORIENTAL HORNET
1992

MEDAKA
1996

CRICKET
1998

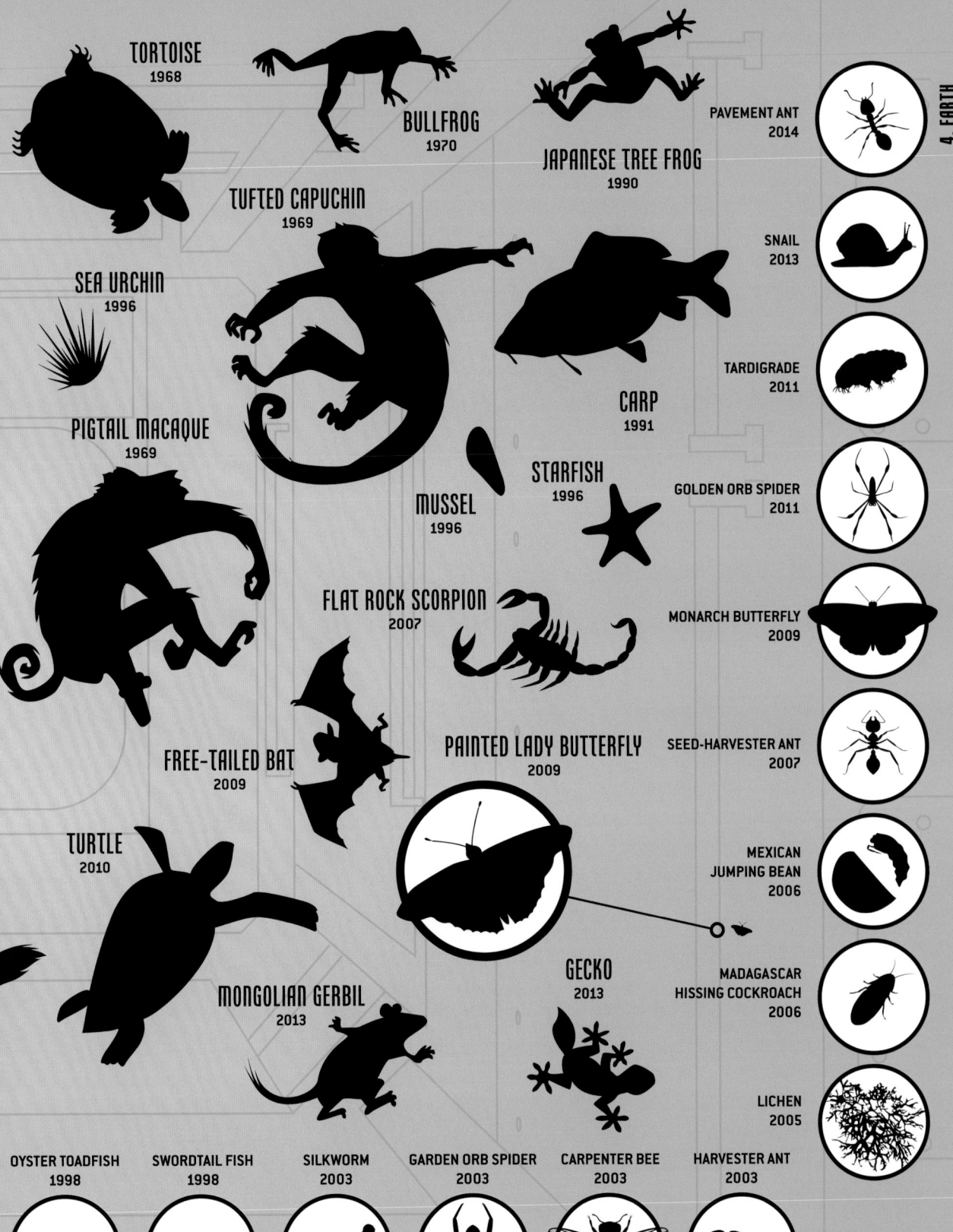

TORTOISE
1968

BULLFROG
1970

JAPANESE TREE FROG
1990

PAVEMENT ANT
2014

TUFTED CAPUCHIN
1969

SNAIL
2013

SEA URCHIN
1996

TARDIGRADE
2011

CARP
1991

PIGTAIL MACAQUE
1969

STARFISH
1996

GOLDEN ORB SPIDER
2011

MUSSEL
1996

FLAT ROCK SCORPION
2007

MONARCH BUTTERFLY
2009

FREE-TAILED BAT
2009

PAINTED LADY BUTTERFLY
2009

SEED-HARVESTER ANT
2007

TURTLE
2010

MEXICAN
JUMPING BEAN
2006

GECKO
2013

MADAGASCAR
HISSING COCKROACH
2006

MONGOLIAN GERBIL
2013

LICHEN
2005

OYSTER TOADFISH
1998

SWORDTAIL FISH
1998

SILKWORM
2003

GARDEN ORB SPIDER
2003

CARPENTER BEE
2003

HARVESTER ANT
2003

4. EARTH

55

PEOPLE IN SPACE – I

Every day that there's been a human being in space.

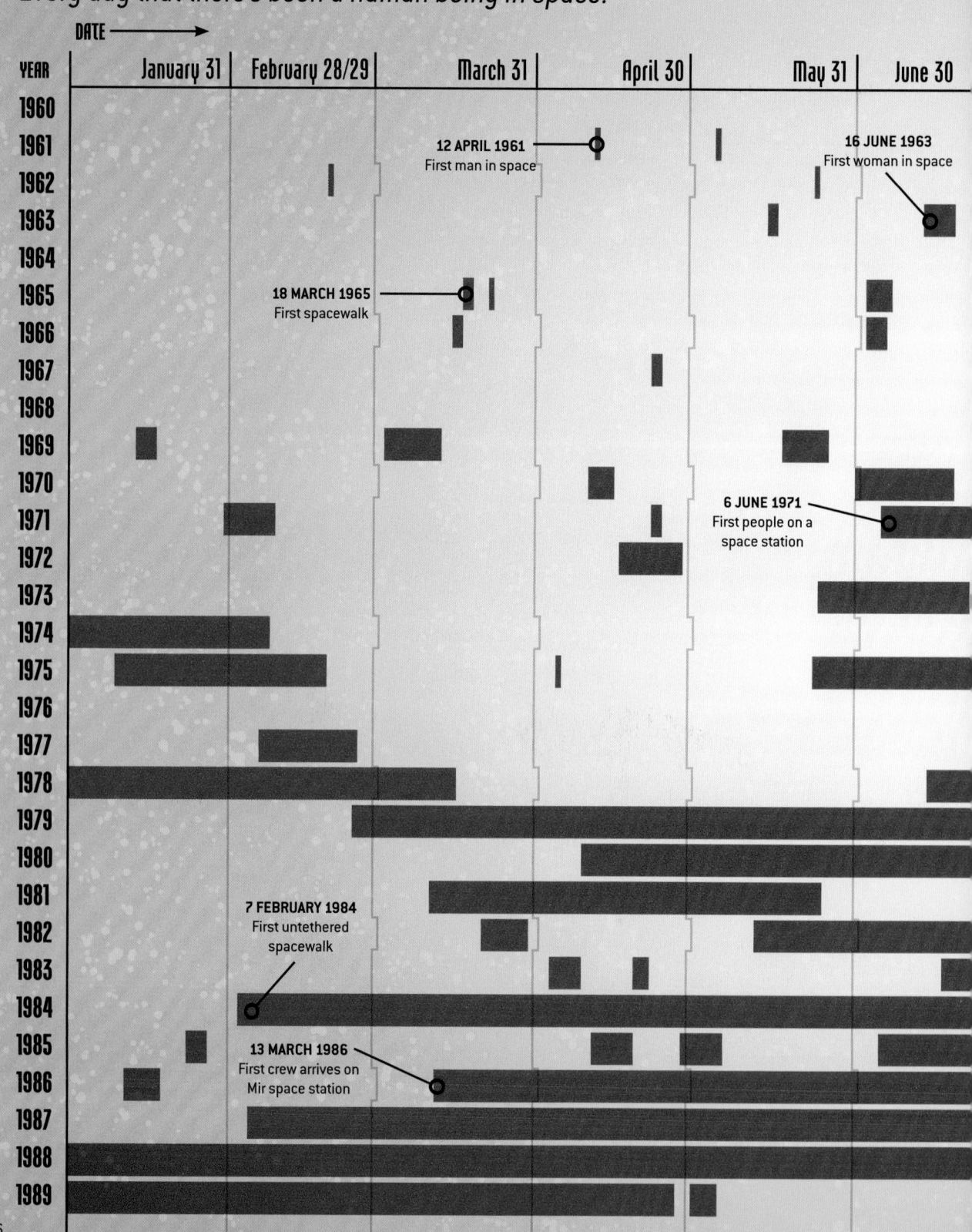

DATE ➞

YEAR	January 31	February 28/29	March 31	April 30	May 31	June 30

12 APRIL 1961
First man in space

16 JUNE 1963
First woman in space

18 MARCH 1965
First spacewalk

6 JUNE 1971
First people on a space station

7 FEBRUARY 1984
First untethered spacewalk

13 MARCH 1986
First crew arrives on Mir space station

KEY

| = 1 day

ADDITIONAL DAY
FOR LEAP YEARS

July 31	August 31	Semptember 30	October 31	November 30	December 31	YEAR
						1960
						1961
						1962
						1963
						1964
						1965
						1966
						1967
						1968
						1969
						1970
						1971
						1972
						1973
						1974
						1975
						1976
						1977
						1978
						1979
						1980
						1981
						1982
						1983
						1984
						1985
						1986
						1987
						1988
						1989

16 JULY 1969
First people on Moon

21 DECEMBER 1968
First people beyond
low-Earth orbit

30 JUNE 1971
First human death in space

14 DECEMBER 1972
Last people on Moon

21 DECEMBER 1988
First people to
complete a whole year
continuously in space

57

PEOPLE IN SPACE – II

Date ⟶

Year	January 31	February 28/29	March 31	April 30	May 31	June
1990						
1991						
1992						
1993						
1994						
1995						
1996						
1997						
1998						
1999						
2000						
2001						
2002						
2003						
2004						
2005						
2006						
2007						
2008						
2009						
2010						
2011						
2012						
2013						
2014						
2015						
2016						
2017						
2018						
2019						
2020+						

16 JUNE 2000
Last crew on Mir space station

28 APRIL 2001
First space "tourist"

19 APRIL 2002
First person to complete 7 trips to space (current record)

21 JUNE 2004
First privately funded human space flight

July 31	August 31	Semptember 30	October 31	November 30	December 31	Year
						1990
						1991
						1992
						1993
						1994
						1995
						1996
						1997
						1998
						1999
						2000
						2001
						2002
						2003
						2004
						2005
						2006
						2007
						2008
						2009
						2010
						2011
						2012
						2013
						2014
						2015
						2016
						2017
						2018
						2019

2 NOV 2000
First crew arrives on International Space Station

17 JULY 2009
13 people on a single craft in space (current record)

18 OCTOBER 2019 First all-woman spacewalk

2020+ There have been people in space ever since

SPACE HAS SIDE EFFECTS

How being in space affects the human body.

INSOMNIA

The body's circadian rhythms are upset in space — for example, on the International Space Station astronauts experience a sunrise and sunset every 45 minutes. Also, space vehicles and stations tend to be noisy and busy. Even with sleeping pills, astronauts often find sleep difficult.

RADIATION

Without the protection of Earth's atmosphere and magnetic fields, astronauts in space are exposed to 10x the amount of radiation as people on Earth — with increasing risks of related diseases such as cancer.

SUNBURN

Exposure to direct, unfiltered sunlight.

LOW PRESSURE

Lack of atmospheric pressure can cause ebullism, when bubbles form in bodily fluids — which can cause serious damage to the body and even death.

G-FORCE

Leaving and returning to Earth, astronauts experience several times the normal force of gravity. At 4–6x normal gravity, people lose consciousness — deadly if they need to operate controls!

BACTERIA

It has been shown that bacteria in space are more resistant to antibiotics and thrive better.

STRESS

The physical effects of being in space, and the isolation and stress associated with the work can cause difficulties for astronauts.

EMISSIONS

Weightlessness means sneezes, burps and going to the toilet can be messy. Escaping material floats around and can make astronauts ill or damage equipment!

SPACE ADAPTATION SYNDROME

Aka "space sickness": headaches, nausea, vomiting and lethargy due to the body being disoriented in micro gravity. It usually affects astronauts in their first few days in space.

EYE CRUSH

Weightlessness increases pressure on the back of the eyeball — changing its shape and damaging the optic nerve, making vision blurry.

EYE FLASH

Outside Earth's magnetic fields, astronauts are exposed to many more cosmic rays with an affect like a sudden flash of light.

"MOON FACE"

Lack of gravity means bodily fluids not drained downwards, leaving astronauts puffy-headed — which they call "Moon face". This affects balance, vision, taste and smell.

LACK OF OXYGEN

In the vacuum of space, astronauts are at risk of asphyxiation and other conditions such as hypoxia.

SPINAL DISC HERNIATION

Many astronauts suffer back pain and even herniated discs after returning to Earth from space, for reasons that are not yet readily understood.

LONG-TERM EFFECTS OF WEIGHTLESSNESS

Include: loss of calcium in bones causing osteoporosis; reduction of muscle mass and aerobic capacity; weakening of heart.

05. MOON

CRUST

* Silicate rocks
* 60 km thick on near side
* 100 km thick on far side

MANTLE

* Silicate rock, probably Peridotite (similar to Earth's mantle)

MARIA (lunar lowlands)

* Smooth, darker regions once thought to be seas
* They're really solidified basalt lava
* Eruptions 3.9–3.6 billion years ago
* (Some volcanism continued until 1.2 billion years ago)
* Basalt about 1 km thick

NO ATMOSPHERE

* Surface has no protection from meteorites or solar radiation

CORE

* Extremely small — about 350 km diameter. Partially liquid iron, nickel
* The spinning of the Moon's core does not produce magnetic fields — as the Earth's core does. However, studies of lunar rock collected by the crew of Apollo 17 in 1972 suggest the Moon had magnetic fields 4.2 billion years ago

SURFACE GRAVITY

* 1/6 of Earth

Side of Moon always facing Earth

Side of Moon unobservable from Earth's surface

RAYS

* "Splash" pattern of ejected material from impacts

CRATERS

* Created by meteorite impacts

SURFACE TEMPERATURE

* Lack of atmosphere and slow rotation = huge range in surface temperature, from −170°C to +130°C

TERRAE (lunar highlands)

* Heavily cratered, lighter regions
* Mountains rise 5 km above surrounding surface
* Composed of anothorsite (igneous rock)
* About 4.5–4 billion years old

REGOLITH

* 'Soil' of rock particles and mineral grains hundreds of metres thick, formed by bombardment of meteorites and solar wind

 EARTH ←

KEY FACTS

Diameter: 3,476 km
Average distance from Earth: 384,400 km
Rotation period: 27.3 days (see side-bar)

Light from the Moon takes **1 second** to reach Earth

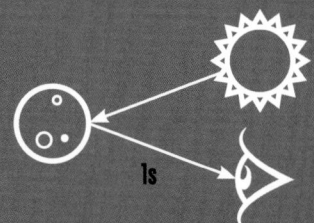

SCALE

EARTH   MOON

FAMILIAR FACE

LUNAR DAY

A lunar day – when it rotates once on its axis – is 27.3 days.

27.3 DAYS

SIDEREAL MONTH

The Moon also takes 27.3 days to orbit the Earth, measured by its position against background stars – a **SIDEREAL** month.

27.3 DAYS

The effect is that it always presents the same side to us.

Although we only see one hemisphere of the Moon, over time we can see 59% of the lunar surface. This is due to the Moon's orbit being elliptical and tilted, and to the width of the Earth enabling us to get different perspectives. This effect is called **LIBRATION**.

SYNODIC MONTH

It takes 29.5 days to see the same phase of the Moon because the Earth is moving around the Sun at the same time. This is a **SYNODIC** month.

SNAP SNAP!

At 3.30 am on 7 October 1959, spacecraft Luna 3 took the first photograph of the far side of Moon.

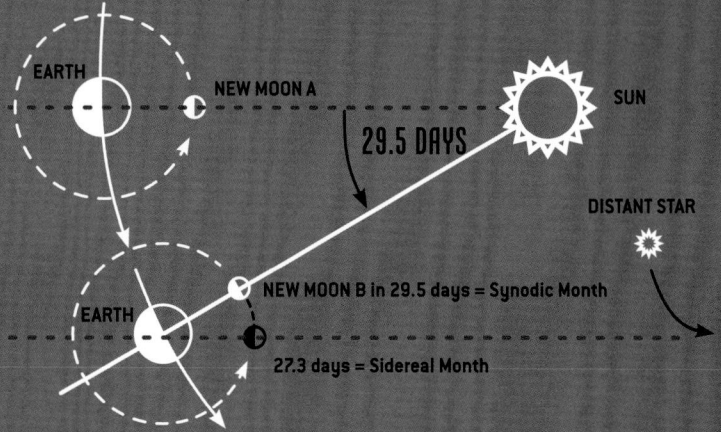

EARTH   NEW MOON A   SUN

29.5 DAYS

DISTANT STAR

NEW MOON B in 29.5 days = Synodic Month

EARTH

27.3 days = Sidereal Month

384,400 KM Distance to scale

MOON ●

ORIGINS

A leading explanation for how the Moon formed is sometimes called the "Big Splash" theory.

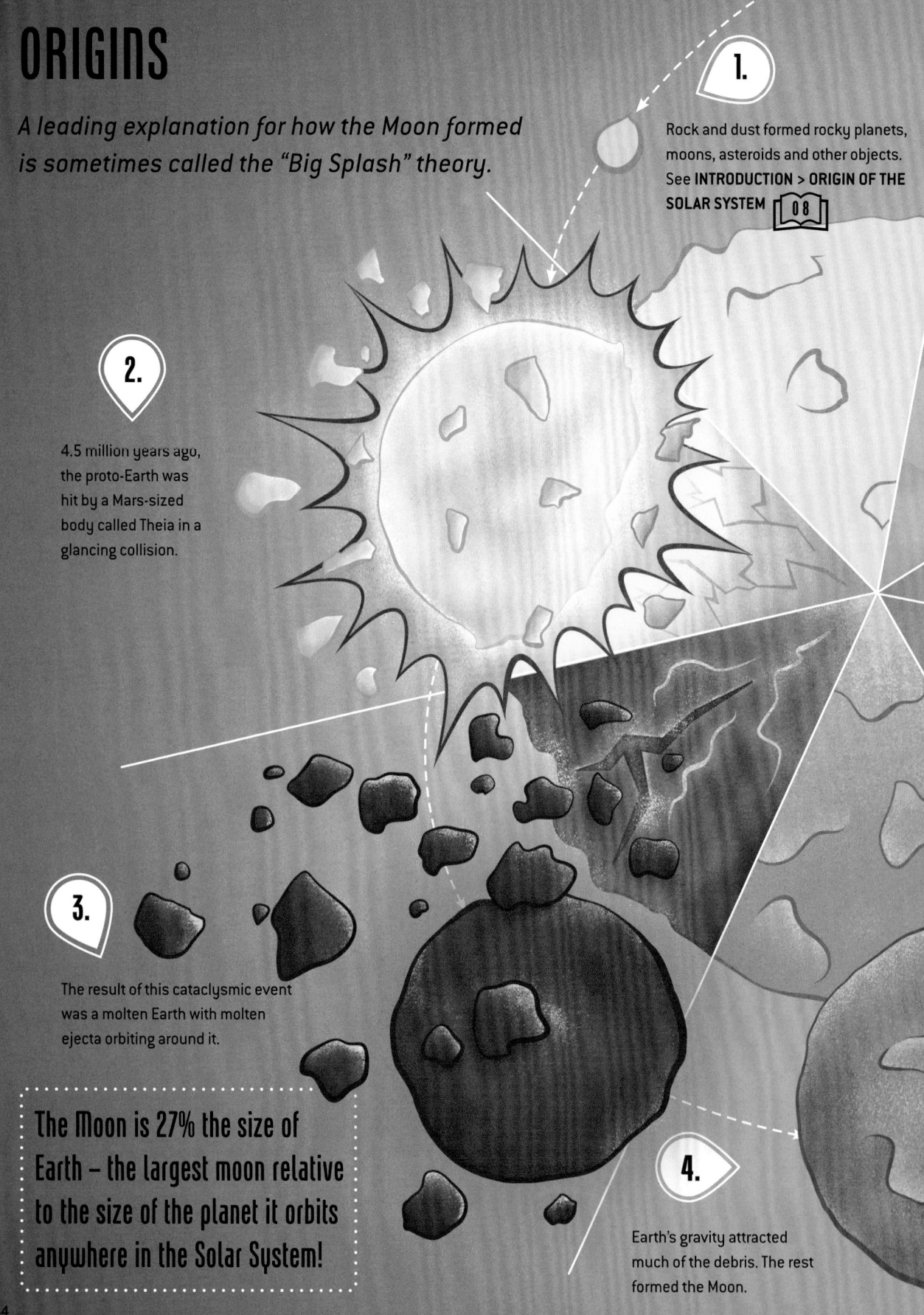

1.

Rock and dust formed rocky planets, moons, asteroids and other objects. See **INTRODUCTION > ORIGIN OF THE SOLAR SYSTEM** `08`

2.

4.5 million years ago, the proto-Earth was hit by a Mars-sized body called Theia in a glancing collision.

3.

The result of this cataclysmic event was a molten Earth with molten ejecta orbiting around it.

The Moon is 27% the size of Earth – the largest moon relative to the size of the planet it orbits anywhere in the Solar System!

4.

Earth's gravity attracted much of the debris. The rest formed the Moon.

7. Volcanic activity is thought to have ended 1.2 million years ago. The face of the Moon has remained largely unchanged since, apart from meteor impacts over the surface.

6. 3.9–3.6 million years ago, volcanic activity created the dark lunar "mares" of basalt lava.

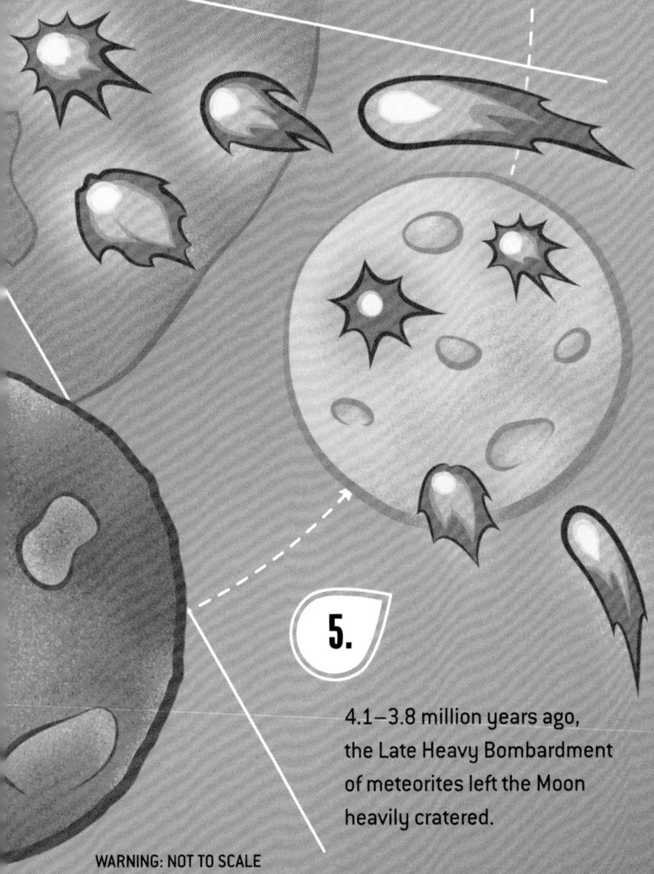

5. 4.1–3.8 million years ago, the Late Heavy Bombardment of meteorites left the Moon heavily cratered.

WARNING: NOT TO SCALE

HOW WE KNOW

There are still many questions about how the Moon formed, but the giant-impact hypothesis is supported by the following evidence:

✳ Isotope ratios of rocks on Earth and the Moon are identical, suggesting shared origin

✳ We see evidence of large collisions elsewhere in the Solar System, and computers can model Earth-Moon collisions

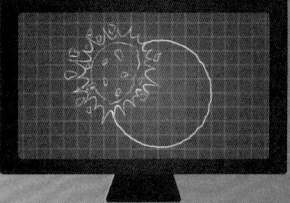

✳ The spin of Earth and orbit of Moon suggest a glancing collision

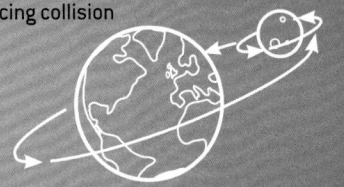

✳ The Moon's small iron core and low density suggest it lost material to Earth's stronger gravity

✳ The Moon is rich in **REFRACTORY** elements with high melting points such as titanium, thorium and uranium, but not in **VOLATILE** elements with low melting points such as sodium, potassium and water. This suggests the Moon's surface was molten. It's thought the volatile elements evaporated in the heat and escaped the Moon's low gravity.

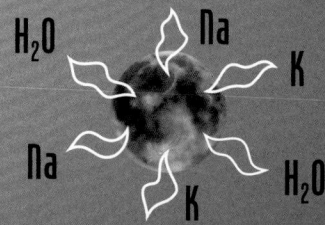

H_2O Na K Na K H_2O

PHASES

Seen from Earth, the appearance of the Moon changes in a regular cycle lasting 29.5 days. Each phase depends on how much of the lit hemisphere of the Moon we can see from Earth.

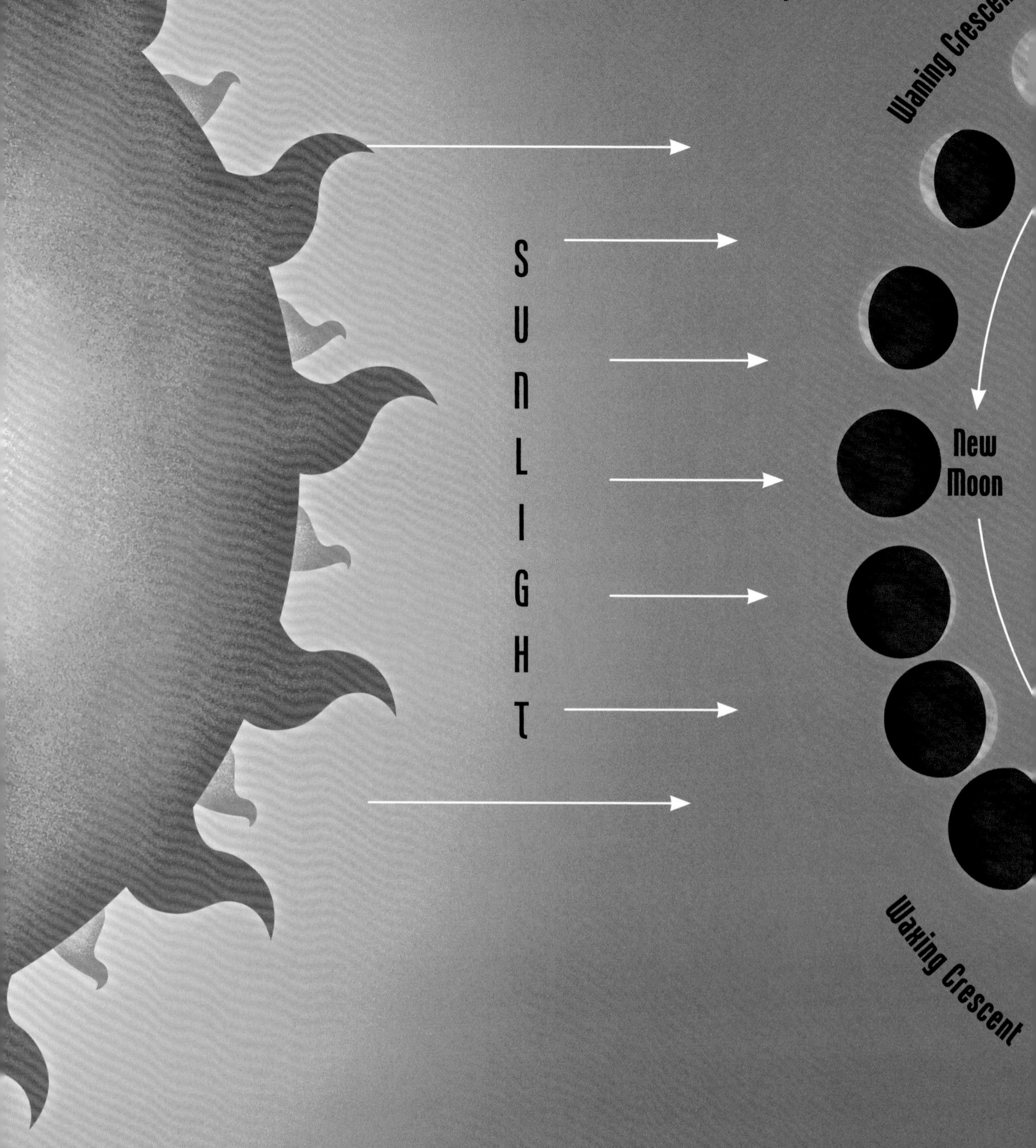

S
U
N
L
I
G
H
T

Waning Crescent

New Moon

Waxing Crescent

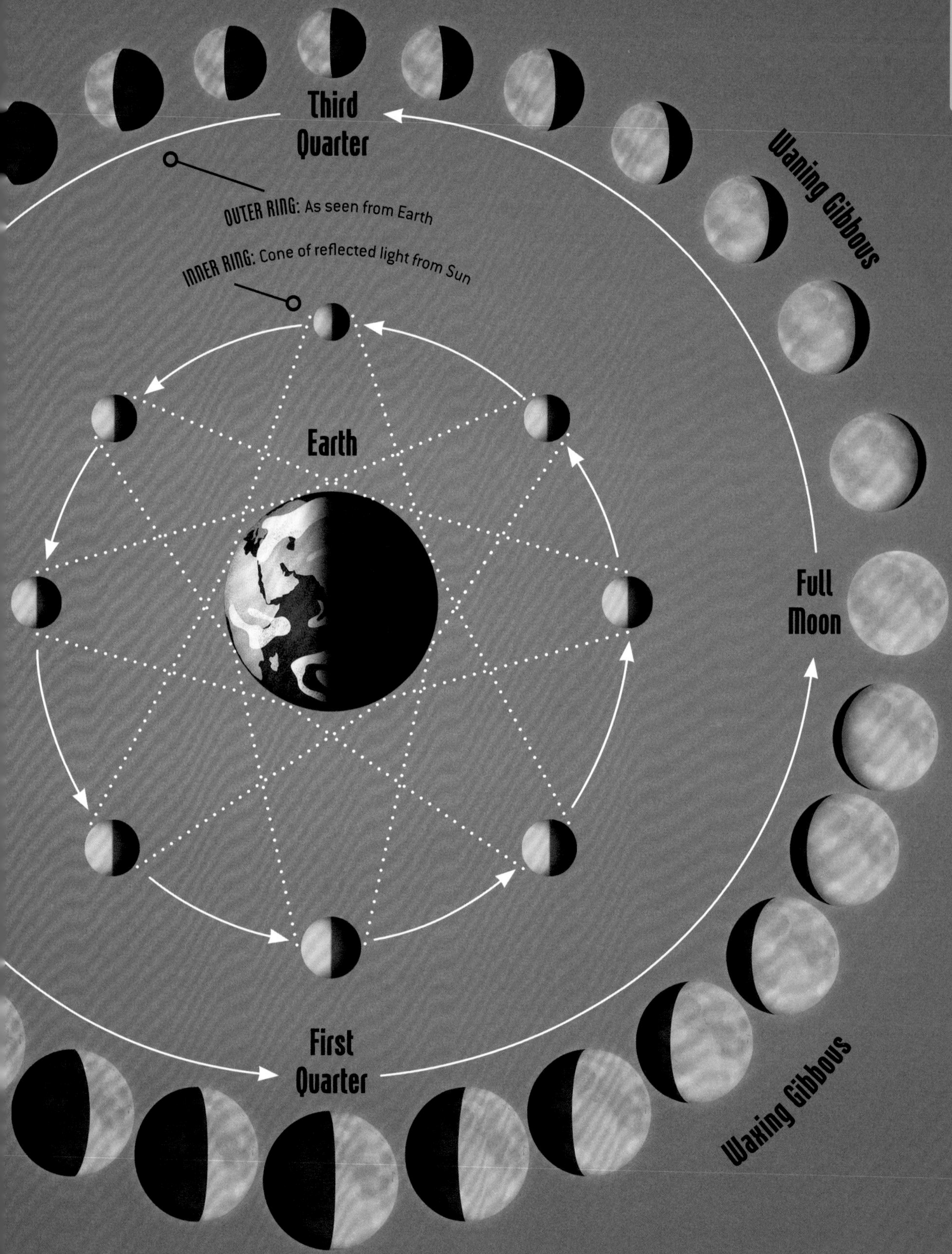

Third
Quarter

OUTER RING: As seen from Earth

INNER RING: Cone of reflected light from Sun

Waning Gibbous

Earth

Full
Moon

First
Quarter

Waxing Gibbous

67

ECLIPSES

When we're in the shadow of the Moon — and vice versa.

SOLAR ECLIPSES

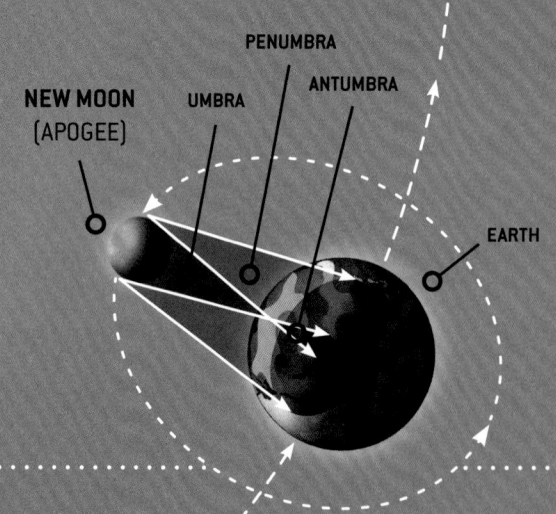

TOTAL ECLIPSE

* Sun = 400x bigger than Moon
* Sun = 400x further away from Earth than Moon
* Viewed from Earth, Sun and Moon seem the same size

SUN

EARTH

NEW MOON
(PERIGREE)
New Moon in path of Sun casts shadow on Earth

UMBRA
Where **UMBRA** (moving spot of shadow) touches the Earth = **REGION OF TOTALITY**, from where we see a **TOTAL ECLIPSE**

PENUMBRA
From within the wider **PENUMBRA**, we see a **PARTIAL ECLIPSE**

ANNULAR ECLIPSE

* The Moon orbits the Earth in an **ELLIPSE**, like a squashed circle
* **APOGEE** = Moon furthest from Earth
* **PERIGEE** = Moon closest to Earth
* When a new Moon is not at perigee in a solar eclipse, the Moon does not completely cover the disc of the Sun, so we see a ring of light or **ANNULUS**

NEW MOON
(APOGEE)

PENUMBRA

UMBRA

ANTUMBRA

EARTH

WHAT WE SEE FROM EARTH

PARTIAL SOLAR ECLIPSE

TOTAL SOLAR ECLIPSE

Moon obscures disc of Sun but we see the Sun's outer atmosphere (corona) and prominences.

ANNULAR SOLAR ECLIPSE

LUNAR ECLIPSES

* Full Moon passes through Earth's shadow

* Earth is bigger than the Moon = casts a bigger shadow = lunar eclipses last longer than solar eclipses and can be seen over more of the Earth

* Sunlight is refracted through our atmosphere — which is why the Moon appears red

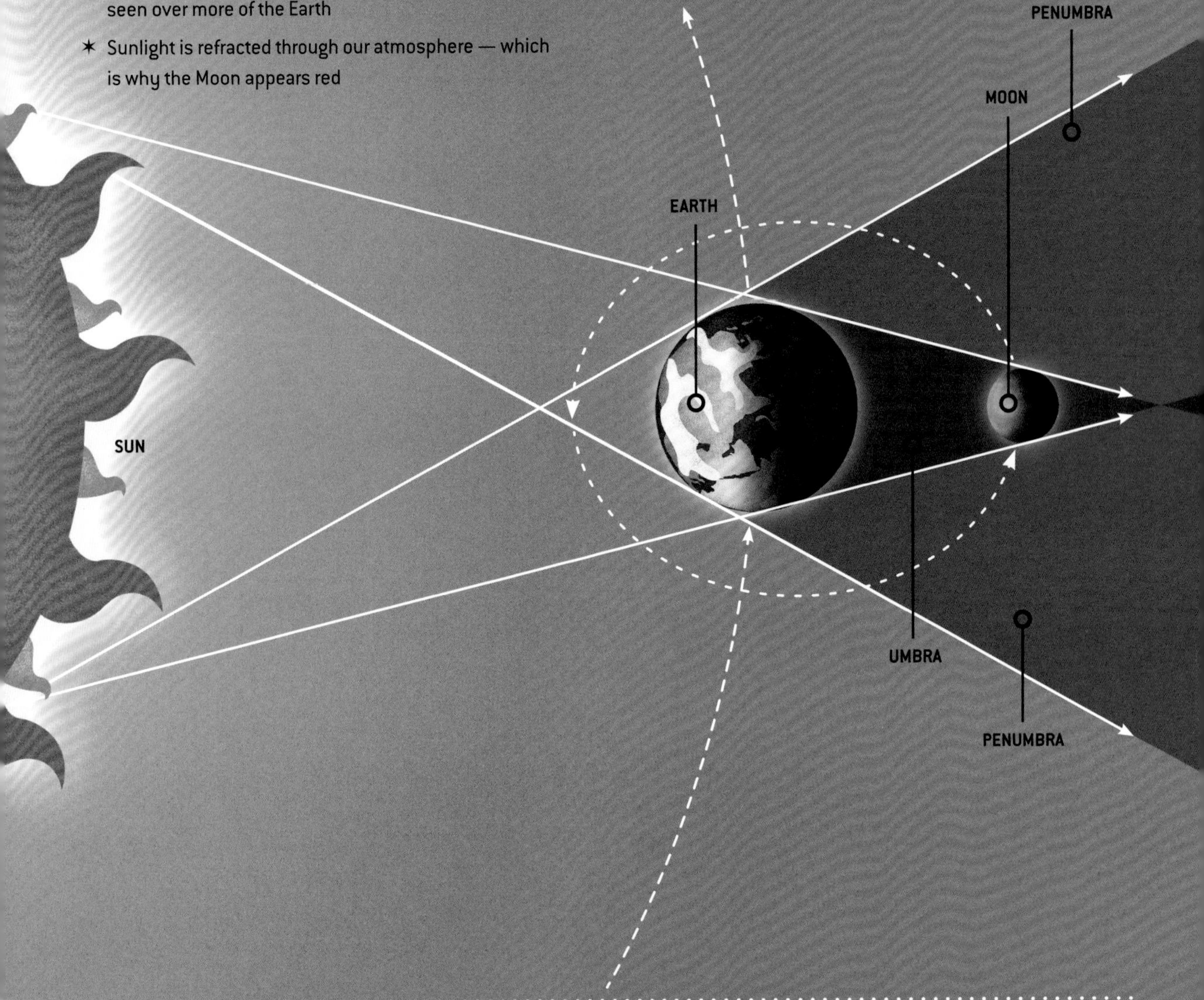

PENUMBRA

MOON

EARTH

SUN

UMBRA

PENUMBRA

WHY AREN'T THERE ECLIPSES EVERY MONTH?

LUNAR ECLIPSE

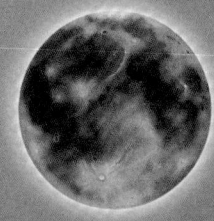

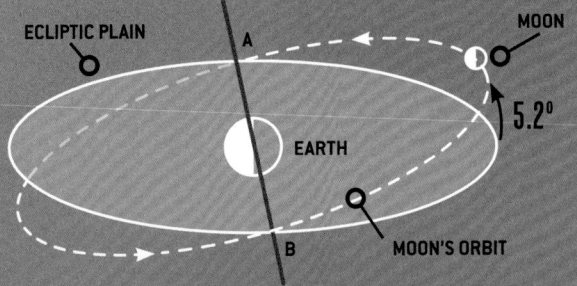

ECLIPTIC PLAIN

A

MOON

5.2°

EARTH

B

MOON'S ORBIT

The plane of the Moon's orbit round Earth is inclined (tilted) 5.2° to that of the ecliptic (the plane of the Earth's orbit around the Sun).

Eclipses occur when a new Moon or full Moon is close to the intersection of the planes — points **A** or **B**.

EXPLORERS

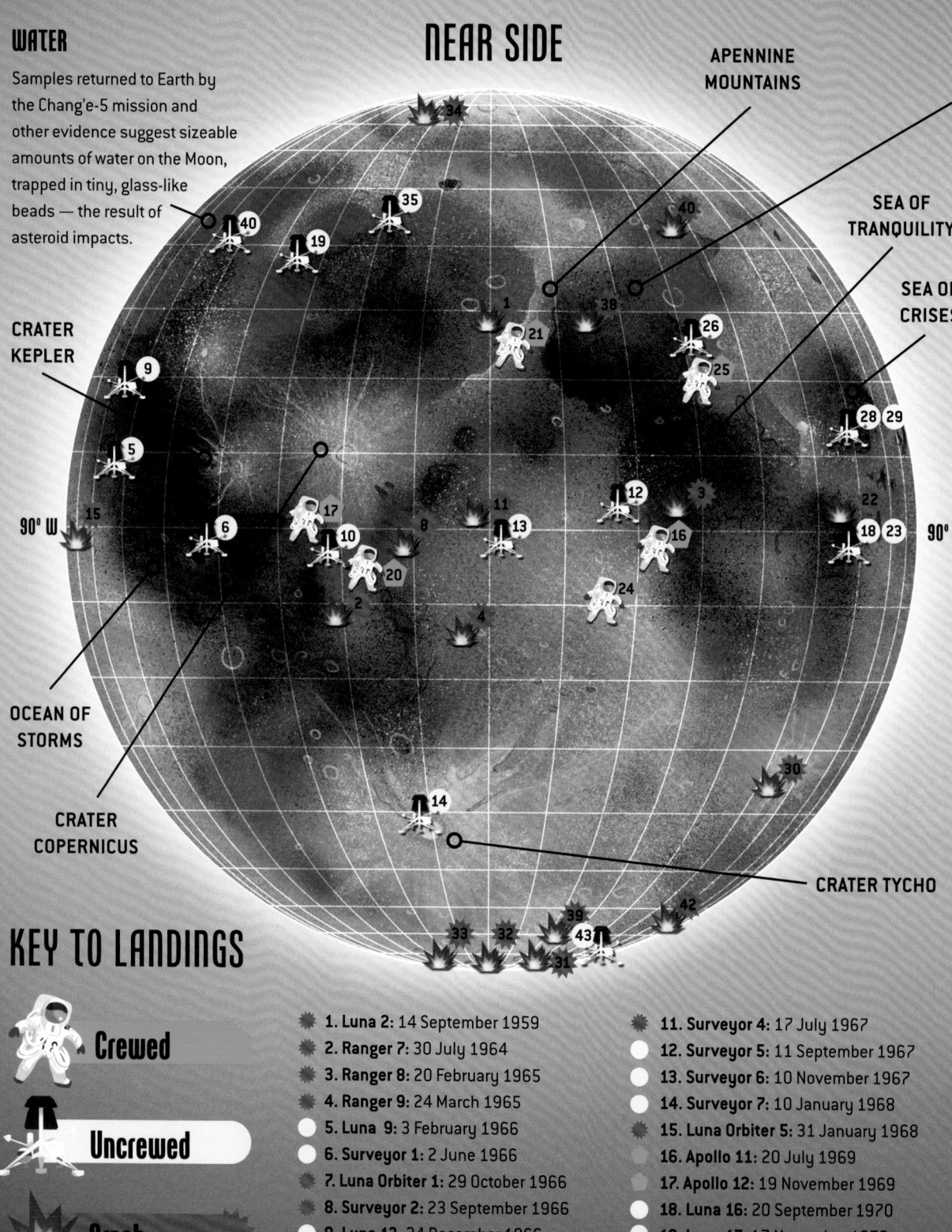

NEAR SIDE

WATER

Samples returned to Earth by the Chang'e-5 mission and other evidence suggest sizeable amounts of water on the Moon, trapped in tiny, glass-like beads — the result of asteroid impacts.

APENNINE MOUNTAINS

SEA OF TRANQUILITY

SEA OF CRISES

CRATER KEPLER

CRATER COPERNICUS

OCEAN OF STORMS

CRATER TYCHO

90° W

90° E

KEY TO LANDINGS

👨‍🚀 **Crewed**

🛰️ **Uncrewed**

💥 **Crash**

- **1. Luna 2:** 14 September 1959
- **2. Ranger 7:** 30 July 1964
- **3. Ranger 8:** 20 February 1965
- **4. Ranger 9:** 24 March 1965
- **5. Luna 9:** 3 February 1966
- **6. Surveyor 1:** 2 June 1966
- **7. Luna Orbiter 1:** 29 October 1966
- **8. Surveyor 2:** 23 September 1966
- **9. Luna 13:** 24 December 1966
- **10. Surveyor 3:** 20 April 1967
- **11. Surveyor 4:** 17 July 1967
- **12. Surveyor 5:** 11 September 1967
- **13. Surveyor 6:** 10 November 1967
- **14. Surveyor 7:** 10 January 1968
- **15. Luna Orbiter 5:** 31 January 1968
- **16. Apollo 11:** 20 July 1969
- **17. Apollo 12:** 19 November 1969
- **18. Luna 16:** 20 September 1970
- **19. Luna 17:** 17 November 1970
- **20. Apollo 14:** 5 February 1971

BERESHEET

Cargo included thousands of **TARDIGRADES**, millimetre-long creatures that survive hostile conditions on Earth. Could they now be living on the Moon?

FAR SIDE

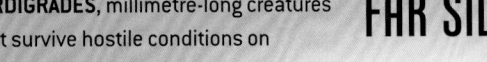

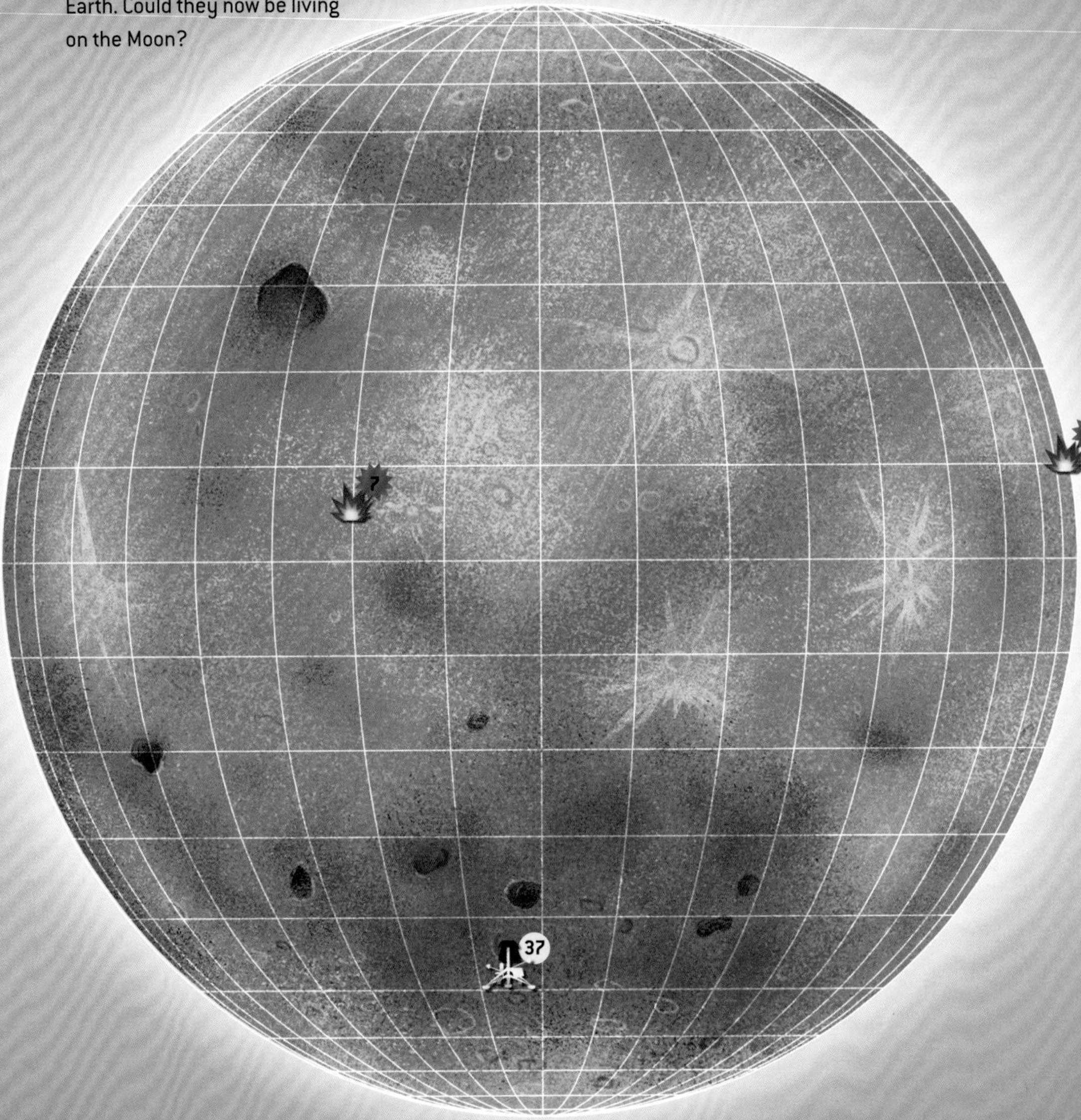

90° E

90° W

21. **Apollo 15:** 30 July 1971
22. **Luna 18:** 11 September 1971
23. **Luna 20:** 21 February 1972
24. **Apollo 16:** 21 April 1972
25. **Apollo 17:** 11 December 1972
26. **Luna 21:** 15 January 1973
27. **Explorer 35:** after 24 June 1973 — landing site unknown
28. **Luna 23:** 6 November 1974
29. **Luna 24:** 18 August 1976

30. **Hiten:** 19 April 1993
31. **Lunar Prospector:** 11 January 1998
32. **Moon Impact Probe:** 14 November 2008
33. **LCROSS:** 9 October 2009
34. **Ebb and Flow:** 17 December 2012
35. **Chang'e 3:** 14 December 2013
36. **LADEE:** 18 April 2014
37. **Chang'e 4:** 3 January 2019
38. **Beresheet:** 11 April 2019
39. **Chandrayaan-2:** 6 September 2019

40. **Chang'e 5:** 1 December 2020
41. **Hakuto-R Mission 1:** 25 April 2023
42. **Luna 25:** 19 August 2023
43. **Chandrayaan-3, Vikram lander:** 23 August 2023
44. **SLIM:** 19 January 2024

LUNAR LAW

The "Outer Space Treaty", ratified in 1967, lays down the dos and don'ts of our visits to the Moon and space in general.

1. NO NATIONAL CLAIMS
— the Moon is for everyone

2. SCIENCE ENCOURAGED

3. EARTH LAWS APPLY
— including the Charter of the United Nations

4. NO WEAPONS OR MILITARY USE

5. HELP EACH OTHER, SHARE INFORMATION
— especially about any dangers

6. EARTH GOVERNMENTS ARE RESPONSIBLE FOR THEIR CITIZENS' ACTIONS ON THE MOON

7. EARTH GOVERNMENTS ARE RESPONSIBLE FOR ANY DAMAGE CAUSED BY THEIR CITIZENS IN GOING TO THE MOON

8. VISITS AND INSPECTIONS WELCOME

9. ACTIVITIES WILL AIM TO ENCOURAGE PEACE AND CO-OPERATION

OTHER LUNAR LAWS

The "Rescue Agreement" (1968) gives guidance on helping other space travellers in emergencies.

The "Moon Agreement" (1979) suggests more detail about property rights etc on the Moon – but has not been widely accepted.

FOR ALL MANKIND

The principles of the "Outer Space Treaty" were summed up by a plaque left behind on the lunar surface by the crew of Apollo 11:

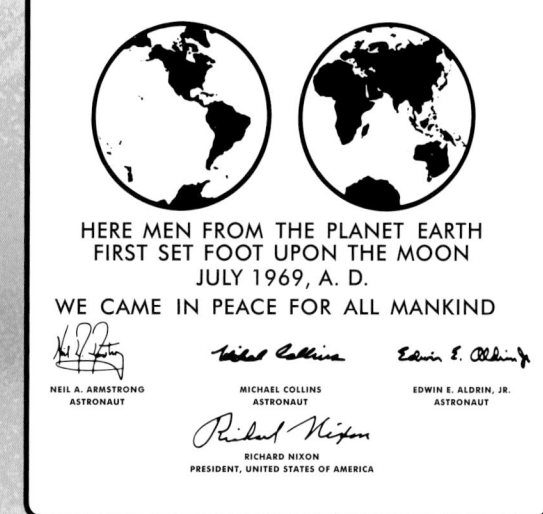

HERE MEN FROM THE PLANET EARTH FIRST SET FOOT UPON THE MOON JULY 1969, A. D.
WE CAME IN PEACE FOR ALL MANKIND

NEIL A. ARMSTRONG
ASTRONAUT

MICHAEL COLLINS
ASTRONAUT

EDWIN E. ALDRIN, JR.
ASTRONAUT

RICHARD NIXON
PRESIDENT, UNITED STATES OF AMERICA

"Here men from the planet Earth first set foot upon the Moon, July 1969 AD. We came in peace for all mankind."

CHEMISTRY

There seems little difference in the composition of the Moon's surface wherever we've looked …

	WATER ICE H_2O	SILICA SiO_2	ALUMINA Al_2O_3
Apollo 11 Samples		42.2%	13.6%
Apollo 12 Samples		46.3%	12.9%
Apollo 14 Samples		48.1%	17.4%
Apollo 15 Samples		46.8%	14.6%
Apollo 16 Samples		45%	27.3%
Apollo 17 Samples		43.2%	17.1%
Luna 16 Samples		41.7%	15.3%
Luna 20 Samples*		45.1%	22.3%
Luna 24 Samples		43.9%	12.5%
All lowlands (maria)		45.4%	14.9%
All highlands (terrae)		45.5%	24%
Shadowed Areas of Polar Craters	3.5% — WATER ICE	96.5% UNKNOWN	

Recently discovered water ice on the Moon excites scientists because it will be very helpful for us living and working there.

THE MOON,
NEAR SIDE

LIME
CaO

IRON (II)
OXIDE
FeO

MAGNESIA
MgO

TITANIUM
DIOXIDE
TiO_2

SODIUM
OXIDE
Na_2O

OTHER

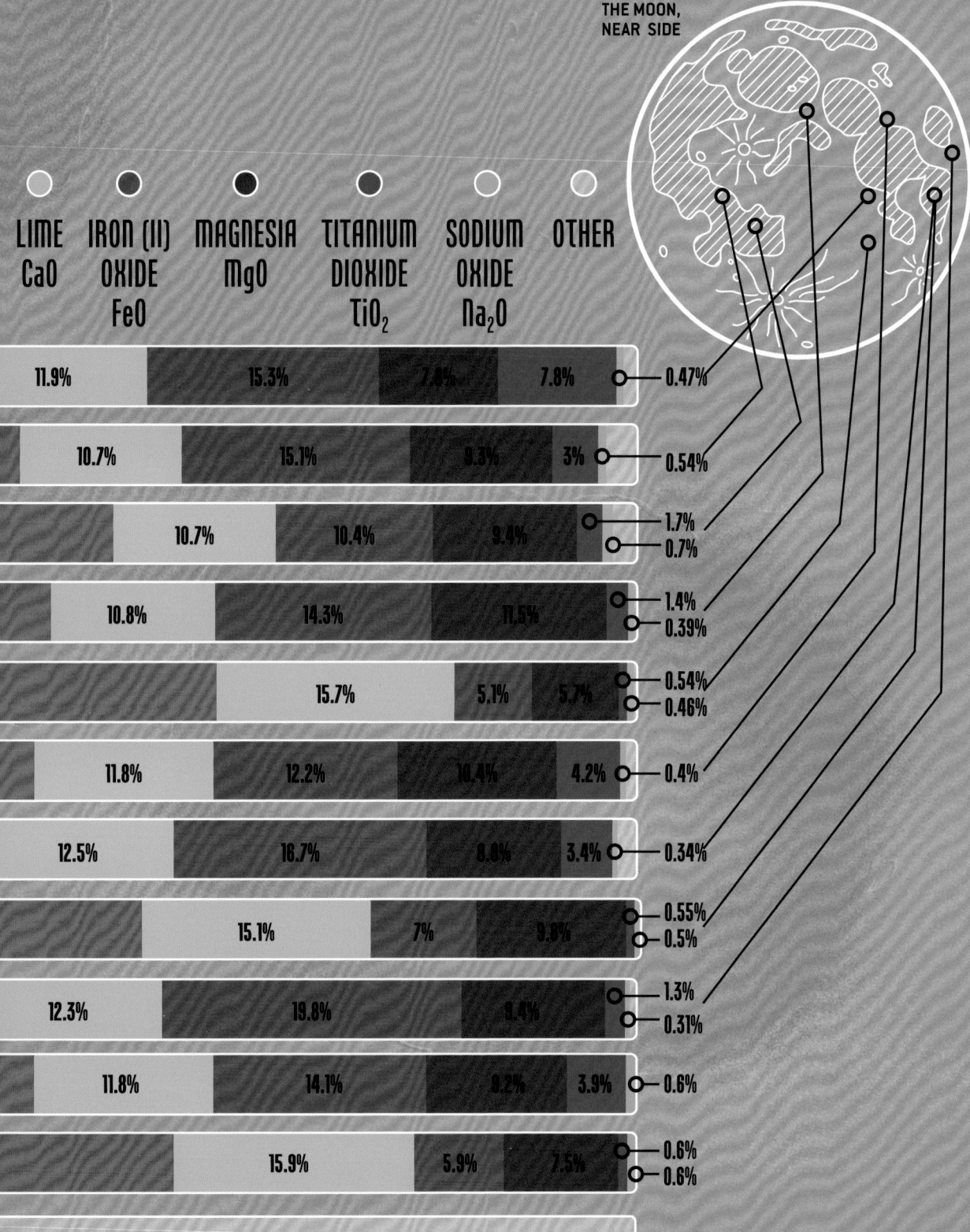

11.9%	15.3%	7.8%	7.8%	0.47%
10.7%	15.1%	9.3%	3%	0.54%
10.7%	10.4%	9.4%		1.7% / 0.7%
10.8%	14.3%	11.5%		1.4% / 0.39%
15.7%	5.1%	5.7%		0.54% / 0.46%
11.8%	12.2%	10.4%	4.2%	0.4%
12.5%	16.7%	8.8%	3.4%	0.34%
15.1%	7%	9.8%		0.55% / 0.5%
12.3%	19.8%	9.4%		1.3% / 0.31%
11.8%	14.1%	8.2%	3.9%	0.6%
15.9%	5.9%	7.5%		0.6% / 0.6%

* The calculation of chemical compositions in weight of average
soils at lunar landing sites can total slightly more or less than 100%

RUNAWAY MOON

How we know that our nearest neighbour is moving away from us by almost 4 cm every year!

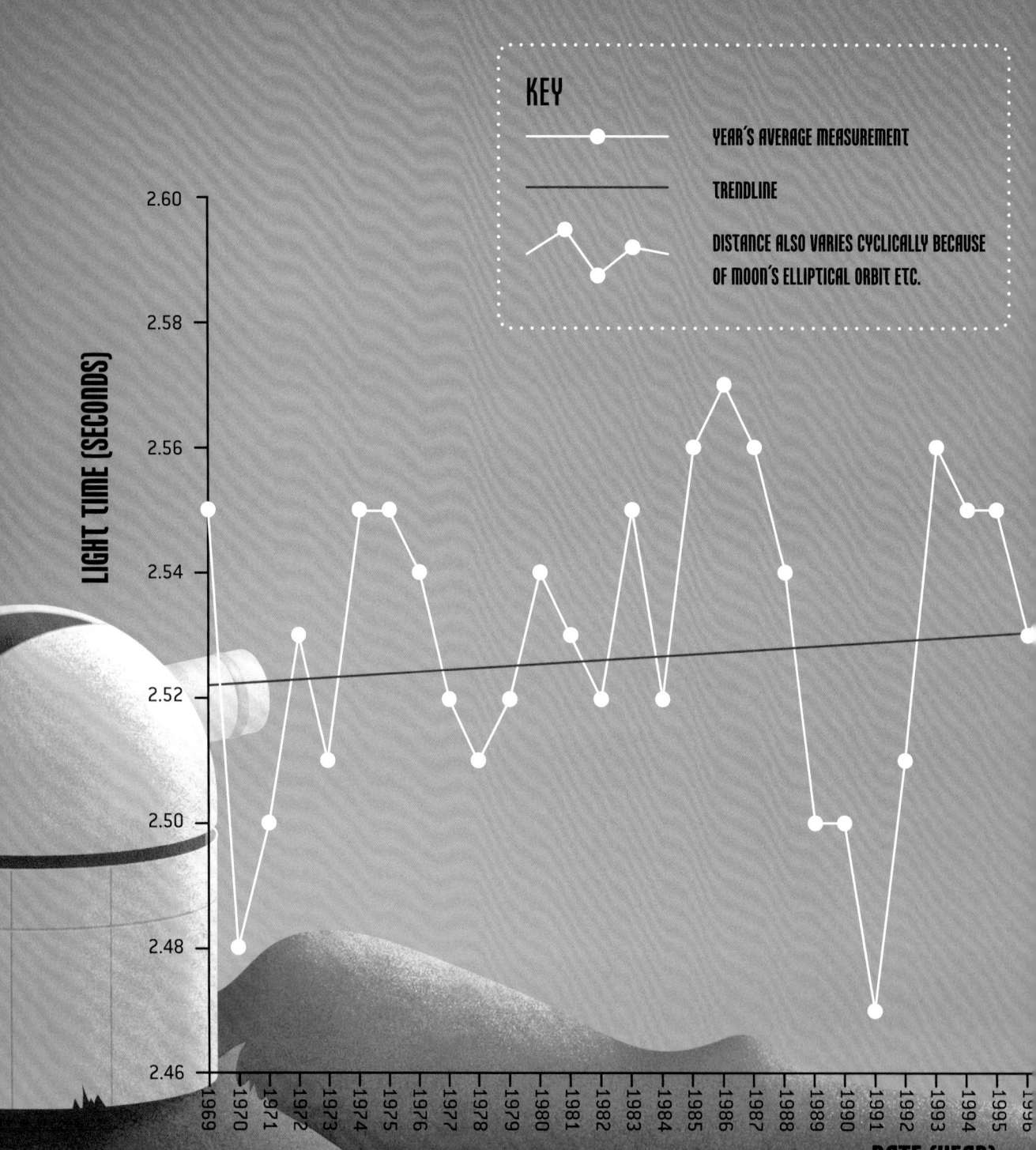

KEY

──●── YEAR'S AVERAGE MEASUREMENT

──── TRENDLINE

⌇⌇ DISTANCE ALSO VARIES CYCLICALLY BECAUSE OF MOON'S ELLIPTICAL ORBIT ETC.

LIGHT TIME (SECONDS)

DATE (YEAR)

Special mirrors called retroreflectors were placed on the lunar surface by Apollo 11 (1969), Luna 17 (1970), Apollo 14 and Apollo 15 (1971), and Luna 21 (1973).

LUNAR LASER RANGING

In July 1969, the first people to walk on the Moon placed the first mirrors there.

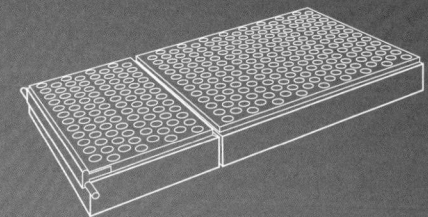

A laser beam fired from Earth at these lunar mirrors is reflected back to us.

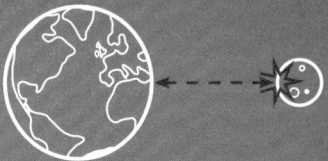

We can measure very accurately how long this return trip takes.

$$2.4956468426s$$

Half that measurement = the time from the Earth to the Moon, one way.

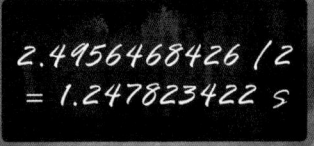

$$2.4956468426 / 2 = 1.247823422 \ s$$

A laser beam travels at the speed of light: **299,792,458 m/s.**

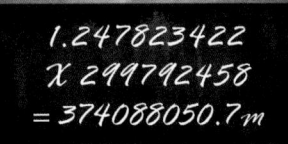

$$1.247823422 \times 299792458 = 374088050.7 m$$

So we can very accurately measure the distance from Earth to the Moon.

Which is how we know the Moon is getting further away from Earth.

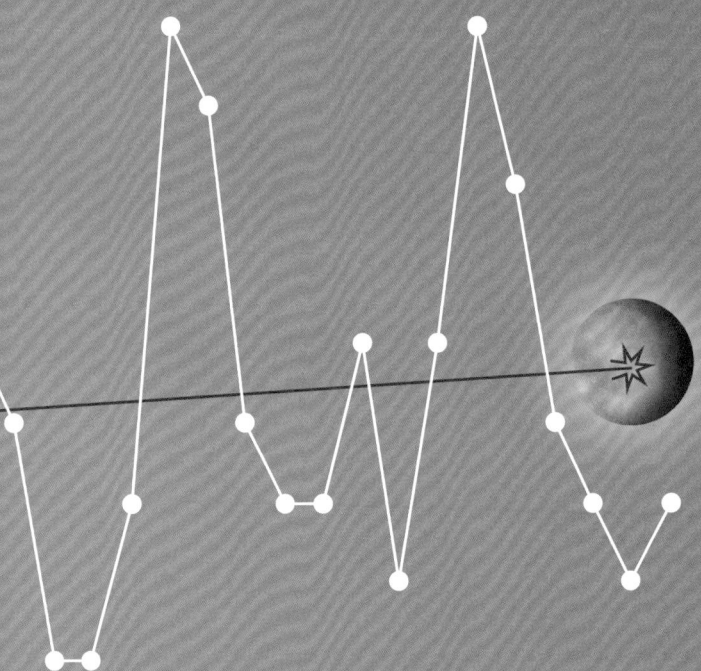

GRAPH BASED ON 51,378 MEASUREMENTS MADE BETWEEN 20 AUGUST 1969 AND 21 DECEMBER 2017, PUBLISHED AT **HTTP://POLAC.OBSPM.FR/LLRDATAE.HTML**

2000 2001 2002 2003 2004 2005 2006 2007 2008 2009 2010 2011 2012 2013 2014 2015 2016 2017

MOON PEOPLE

Just 12 human beings have — briefly — walked on the surface of our nearest neighbour.

APOLLO 11, 1969

TOTAL TIME ON LUNAR SURFACE: 21:35:46

TOTAL EXTRA-VEHICULAR ACTIVITY (EVA): 02:31:40

SAMPLE MASS (KG): 21.55

APOLLO 12, 1969

TOTAL TIME ON LUNAR SURFACE: 31:31:12

TOTAL EVA: 07:45:18

SAMPLE MASS (KG): 34.35

APOLLO 14, 1971

TOTAL TIME ON LUNAR SURFACE: 33:30:31

TOTAL EVA: 09:22:31

SAMPLE MASS (KG): 42.8

NEIL A. ARMSTRONG, EDWIN E. (BUZZ) ALDRIN, JR.

CHARLES CONRAD, JR., ALAN L. BEAN

ALAN B. SHEPARD, JR., EDGAR D. MITCHELL

DAVID R. SCOTT, JAMES B. IRWIN

JOHN W. YOUNG, CHARLES M. DUKE, JR.

EUGENE A. CERNAN, HARRISON H. SCHMITT

KEY

 INTRAVEHICULAR ACTIVITY

 EVA: 1 PAIR = 1 HOUR

 LUNAR ROVING VEHICLE (LRV): 1 = 10 KM

 SAMPLE MASS: 1 = 10 KG

APOLLO 15, 1971

TOTAL TIME ON LUNAR SURFACE: 66:55:04 | **TOTAL EVA:** 19:07:53

DISTANCE COVERED BY ROVER (KM): 27.9

SAMPLE MASS (KG): 77

APOLLO 16, 1972

TOTAL TIME ON LUNAR SURFACE: 71:02:12 | **TOTAL EVA:** 20:14:14

DISTANCE COVERED BY ROVER (KM): 26.7

SAMPLE MASS (KG): 95.71

APOLLO 17, 1972

TOTAL TIME ON LUNAR SURFACE: 74:59:40 | **TOTAL EVA:** 22:03:57

DISTANCE COVERED BY ROVER (KM): 35.74

SAMPLE MASS (KG): 110.52

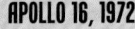
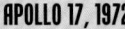

IN LUNAR ORBIT

Apollo 8 — Frank Borman, James A. Lovell,
William A. Anders*
Apollo 10 — Thomas P. Stafford, John W. Young,
Eugene A. Cernan*
Apollo 11 — Michael Collins
Apollo 12 — Richard Gordon
Apollo 13 — James A. Lovell, John L. Swigert, Jr.,
Fred W. Haise, Jr.*
Apollo 14 — Stuart Roosa
Apollo 15 — Alfred Worden
Apollo 16 — Thomas Mattingly
Apollo 17 — Ronald E Evans

*No landing made, orbit only

BONUS NUMBERS

TIME SPENT ON LUNAR SURFACE — 12 D 11 H 34 MIN

TIME SPENT IN EVA — 3 D 9 H 5 MIN 33 S

NO. OF PEOPLE TO SET FOOT ON LUNAR SURFACE — 12

DISTANCE TRAVELLED USING LRV — 90.34 KM

06. TIME AND SPACE

The movements of the Sun, Earth and Moon define our sense of time.

TIME

SECOND ┐
MINUTE ├ Divisions of day
HOUR ┘

DAY — Time taken for Earth to rotate once on its axis

WEEK — In many languages, the names for the days of the week derive from the seven "classical planets" of antiquity (which included the Sun and Moon) and their associated gods

MONTH — Based on time taken for Moon to complete one orbit of Earth

YEAR — Time taken for Earth to complete one orbit of Sun

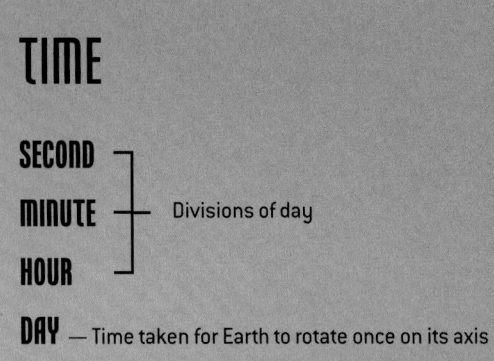

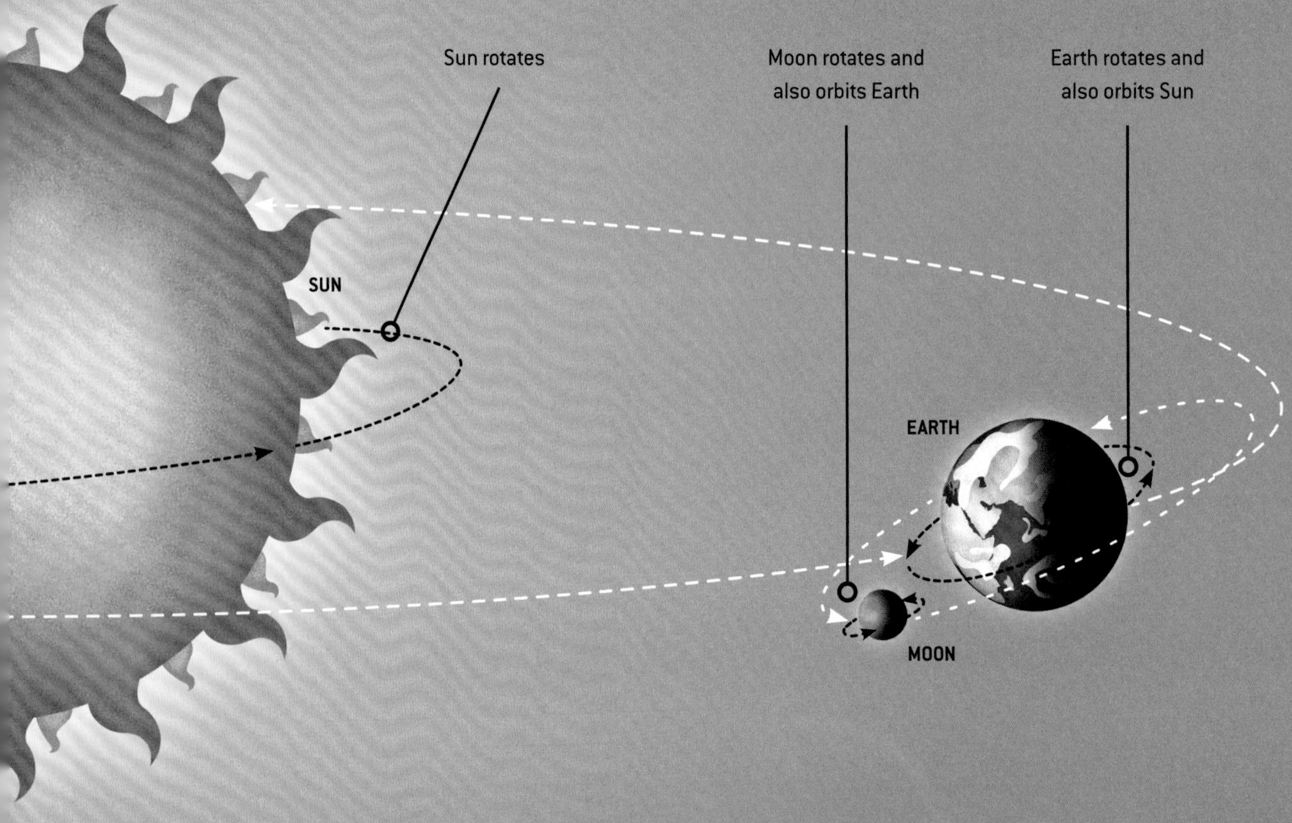

Sun rotates

Moon rotates and also orbits Earth

Earth rotates and also orbits Sun

SUN

EARTH

MOON

Today, we measure time more accurately than by the movement
of celestial bodies, requiring a precise definition:

A second is defined by the time it takes for a specific wavelength of light,
generated by a defined transition in a caesium atom, to complete a very
specific number of oscillations (9,192,631,770 to be precise)

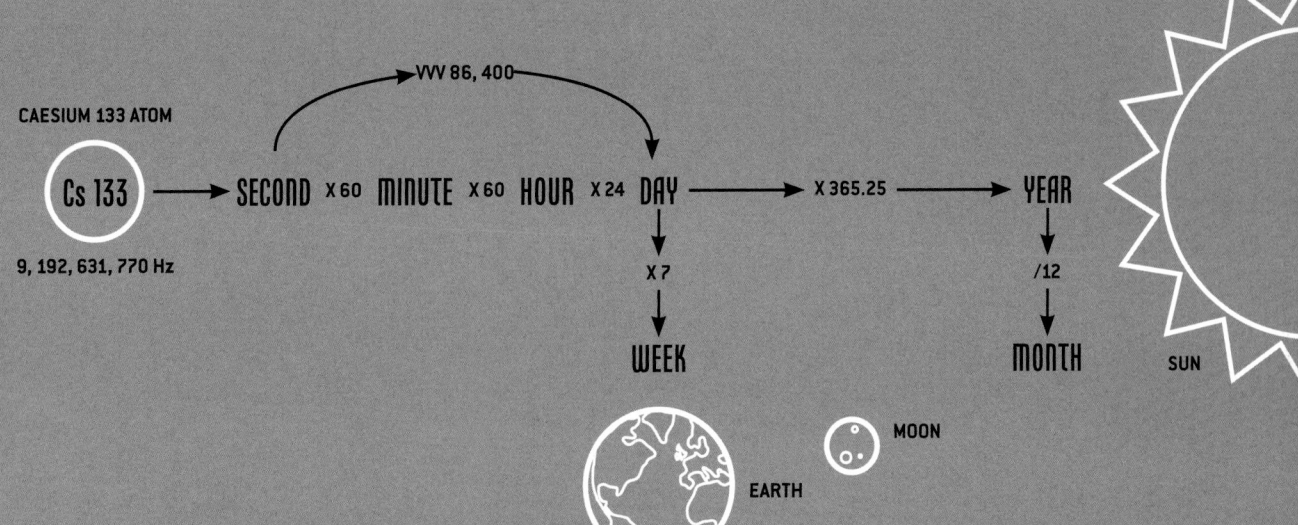

CAESIUM 133 ATOM

VVV 86, 400

Cs 133 → SECOND ×60 MINUTE ×60 HOUR ×24 DAY → ×365.25 → YEAR

9, 192, 631, 770 Hz

DAY ↓ ×7

YEAR ↓ /12

WEEK

MONTH   SUN

EARTH   MOON

OBLIQUITY

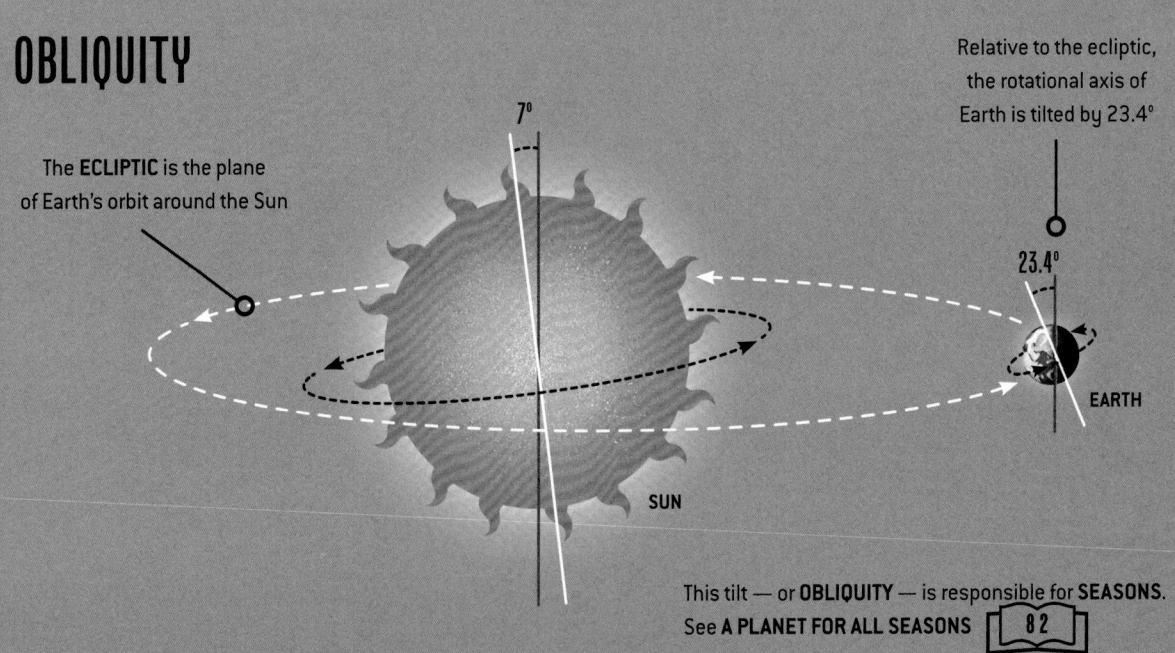

Relative to the ecliptic,
the rotational axis of
Earth is tilted by 23.4°

7°

The **ECLIPTIC** is the plane
of Earth's orbit around the Sun

23.4°

EARTH

SUN

This tilt — or **OBLIQUITY** — is responsible for **SEASONS**.
See **A PLANET FOR ALL SEASONS** 82

A PLANET FOR ALL SEASONS

Seasons on Earth and the other planets are the result of obliquity.

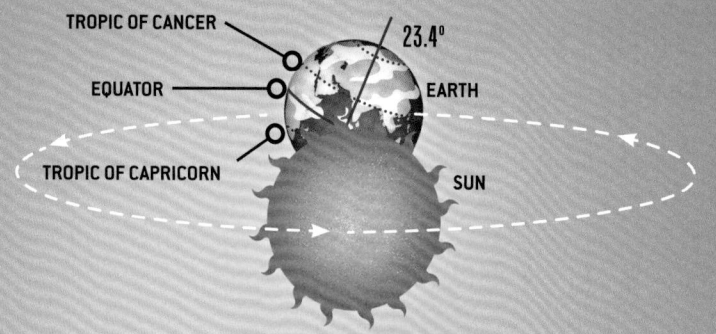

SPRING EQUINOX (~20 MARCH)

* Sun over equator (latitude 0°)
* North and south both have days and nights of 12 hours each
* As seen from Earth, Sun moving north
* Spring in north, autumn in south

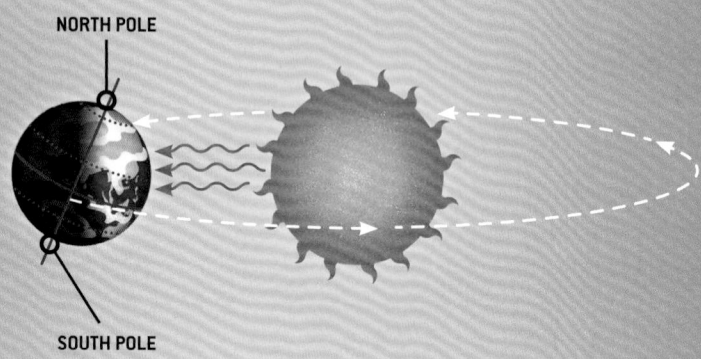

SUMMER SOLSTICE (~21 JUNE)

* In north, Sun highest in sky and day length at maximum
* Sun over Tropic of Cancer (latitude 23°26'N)
* Summer in north, winter in south
* South Pole (90°S) in darkness for six months
* Antarctic circle (66°33'S) in darkness for 24 hours

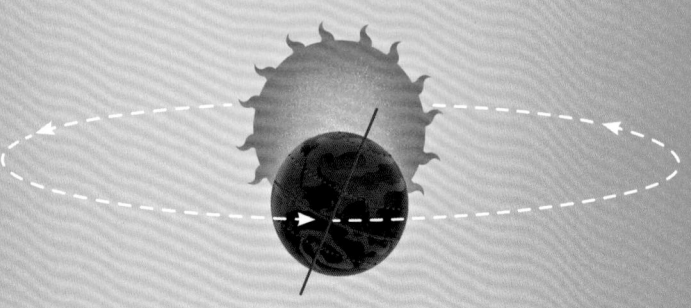

AUTUMN EQUINOX (~22 SEPTEMBER)

* Sun again over equator
* North and south both have days and nights of 12 hours each
* Sun moving south
* Autumn in north, spring in south

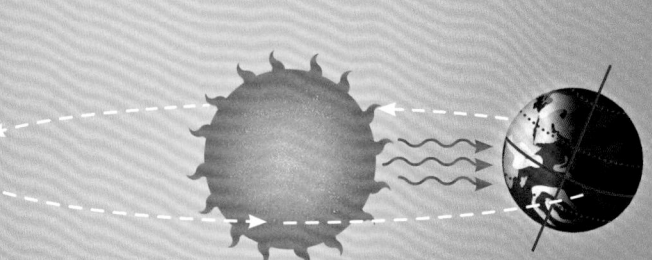

WINTER SOLSTICE (~21 DECEMBER)

* North Pole (90°N) in darkness for six months!
* Arctic Circle (66°33'N) in darkness for 24 hours
* In north, Sun lowest in sky and day length at minimum
* Sun over Tropic of Capricorn (latitude 23°26'S)
* Winter in north, summer in south

WHY IS IT HOTTER IN SUMMER?

In winter, the Sun's radiation strikes the Earth's surface at a shallow angle — and for a shorter time.

In summer, the angle of the Sun's rays is more vertical so more radiation reaches the same area of Earth's surface — and for longer because of the length of the days.

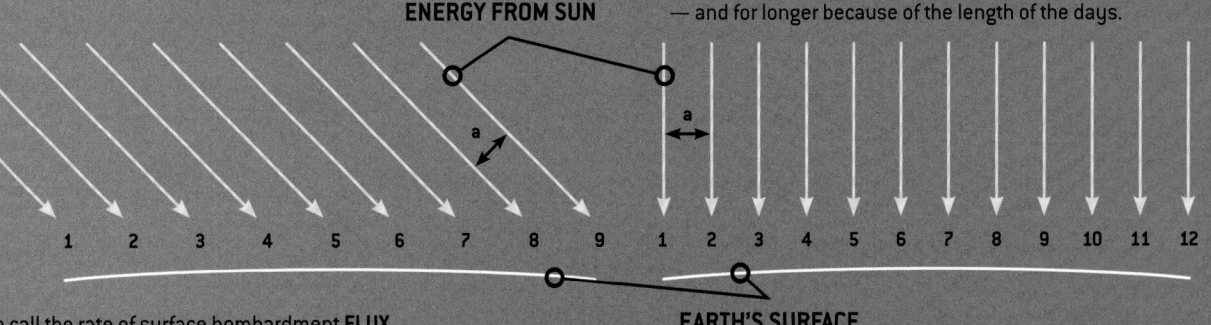

ENERGY FROM SUN

EARTH'S SURFACE

We call the rate of surface bombardment **FLUX**.

SEASONS ON OTHER PLANETS

Mars, Saturn and Neptune have obliquities like Earth and similar patterns of seasons.

At close to 180° obliquity, Venus is "upside down" compared to other planets but otherwise the effect is like have an obliquity of 0° — so it doesn't have seasons.

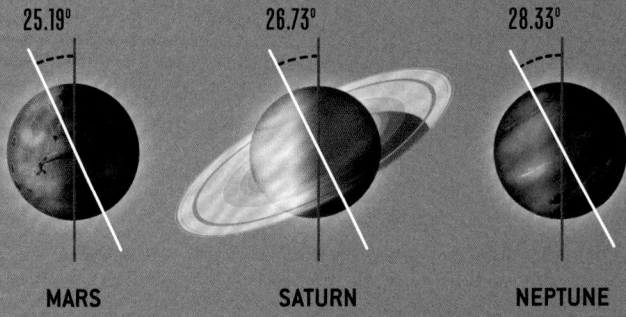

25.19° 26.73° 28.33°

MARS   **SATURN**   **NEPTUNE**

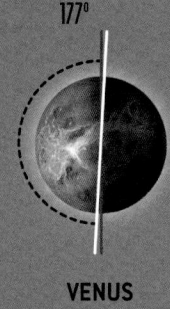

177°

VENUS

Mercury and Jupiter have obliquities near 0° and no seasons.

For half of a Uranus year — that is, 42 Earth years! — one pole points towards the Sun and is in continuous daylight, while the other pole is in continual night!

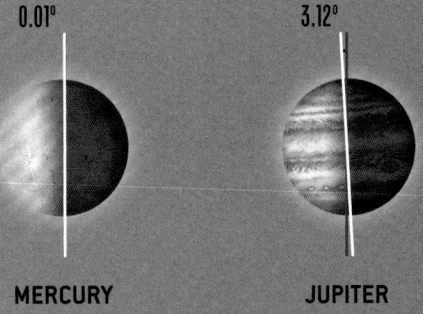

0.01°   3.12°

MERCURY   **JUPITER**

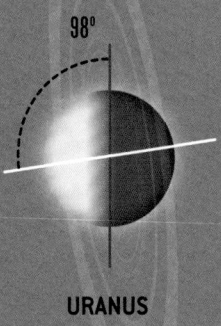

98°

URANUS

VARIATIONS IN DAYLIGHT

Our planet is spinning and tilted, which affects the appearance of the Sun in the sky.

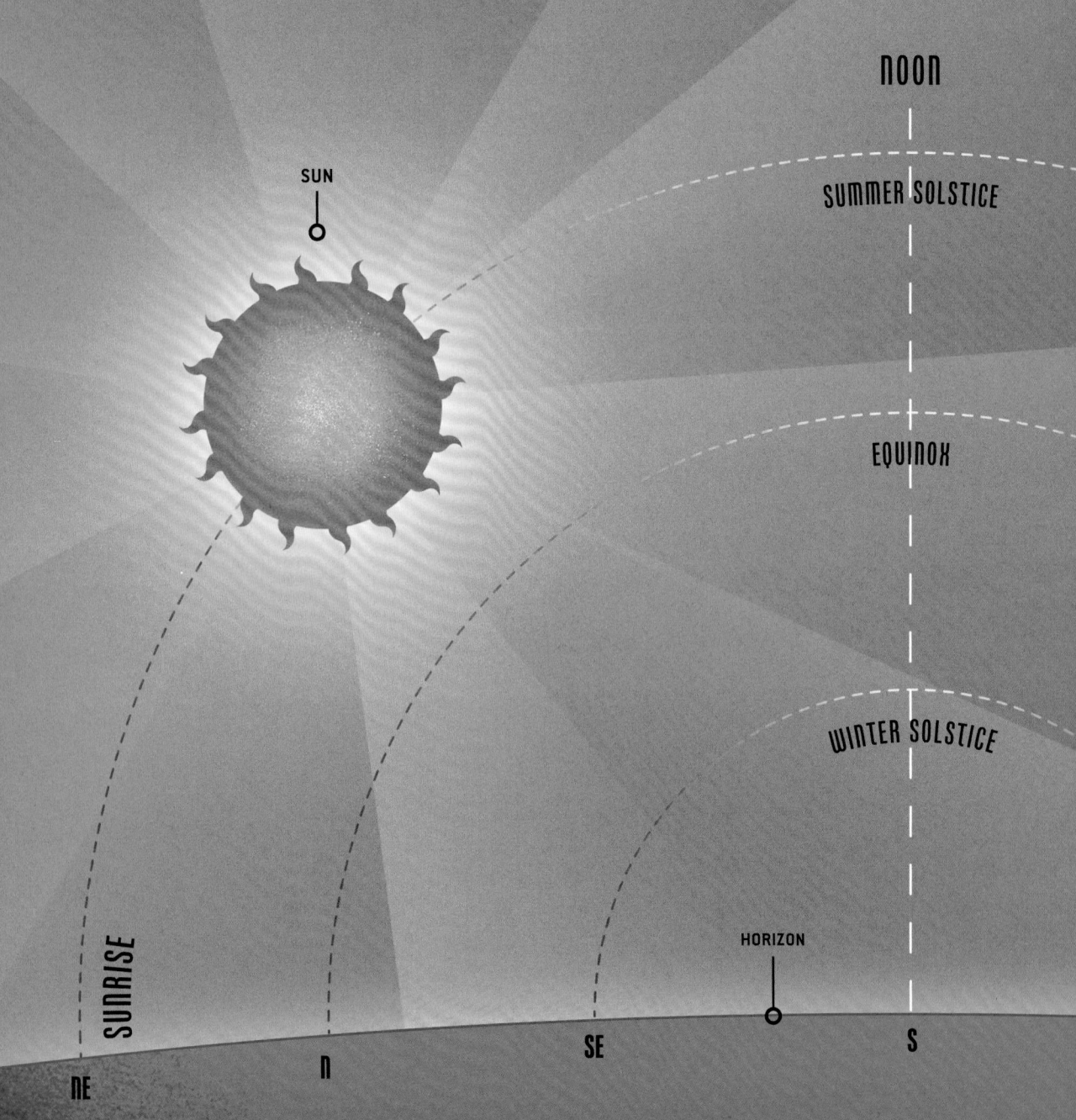

SUN

NOON

SUMMER SOLSTICE

EQUINOX

WINTER SOLSTICE

HORIZON

SUNRISE

NE

N

SE

S

This affects the number of hours of daylight, as does your latitude on Earth.

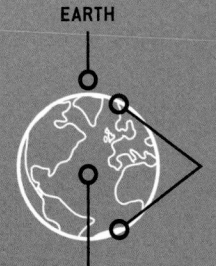

EARTH

LONG DAYS IN SUMMER,
SHORT DAYS IN WINTER

LITTLE VARIATION IN
DAYLIGHT OVER YEAR

VARIATIONS IN SHADOW

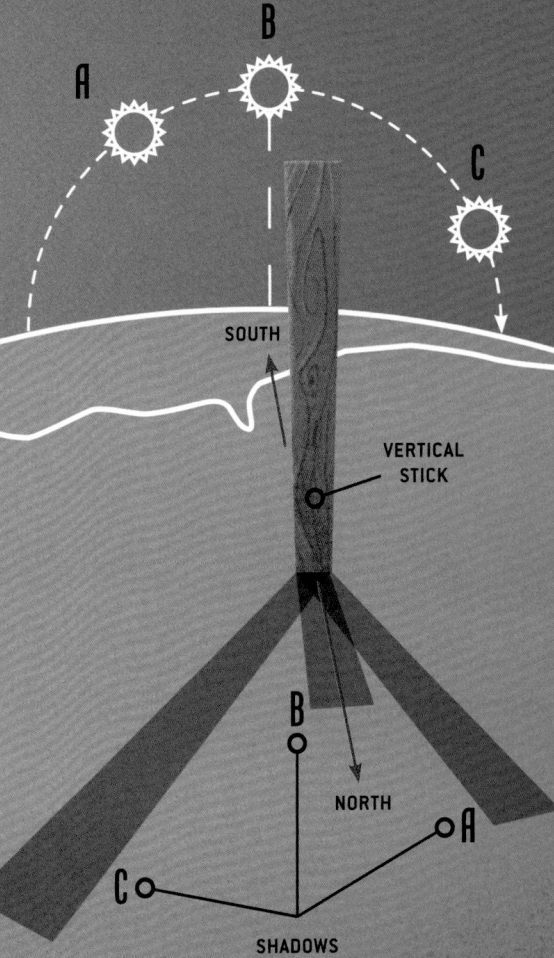

A
B
C

SOUTH

VERTICAL
STICK

B

NORTH

A

C

SHADOWS

APPARENT NOON

Apparent noon is when the shadow is at its shortest, and the Sun is highest in the sky.

At noon, the shadow points true north.

SUNSET

SW

W

NW

HOW TO TELL TIME

When using the Sun to tell the time, we must factor in the movement of Earth.

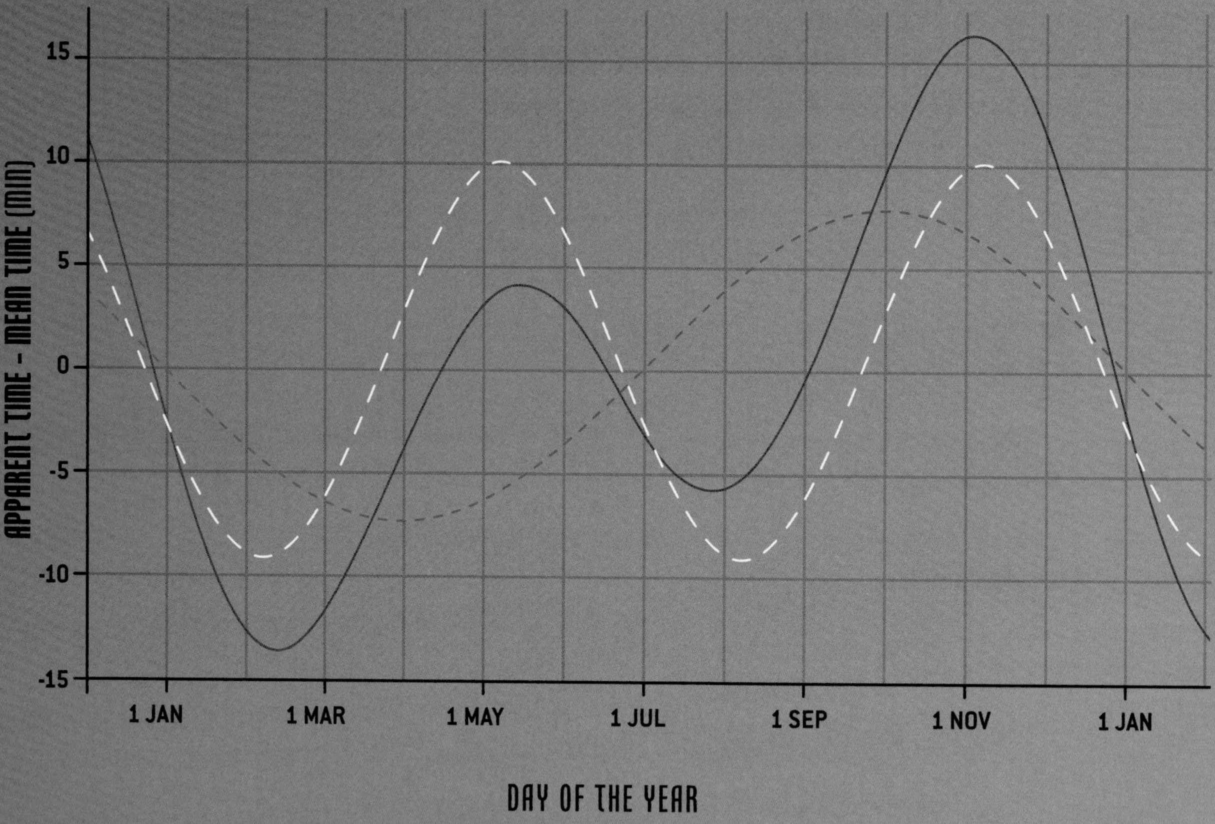

APPARENT TIME – MEAN TIME (MIN)

15

10

5

0

-5

-10

-15

1 JAN   1 MAR   1 MAY   1 JUL   1 SEP   1 NOV   1 JAN

DAY OF THE YEAR

We can use the Sun to measure time in two ways:

SUNDIAL

Time indicated by a **SUNDIAL** = "apparent solar time".

But because of the movement of Earth, daylength varies throughout a year.

CLOCKS

Time given by **CLOCKS** = an average day length of 24 h, so that each day is the same length = "mean solar time".

Apparent solar time – mean solar time = "the **EQUATION OF TIME**", which can be used to convert one to the other.

KEY

———— EQUATION OF TIME, COMPOSED OF:

– – – – VARIATION DUE TO OBLIQUITY (TILT) OF EARTH

- - - - - - VARIATION DUE TO EARTH'S ORBIT ROUND SUN NOT BEING A PERFECT CIRCLE

ANALEMMA

As a result of the Earth's movement, the position of the Sun at the same mean solar time each day changes over the year, in a figure-of-eight pattern known as an "analemma".

SUMMER SOLSTICE

WINTER SOLSTICE

EXAMPLE

On 14 October, the Equation of Time = +14 min.

An accurate sundial at Greenwich (0° longitude) shows 12:14pm apparent solar time.

So the mean solar time = 12:14 – 14 min = 12 noon.

The spin of the Earth on its axis means we must + or – an extra 4 minutes for every degree of longitude from 0°

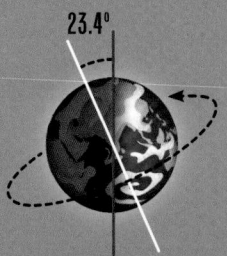

23.4°

THE RIGHT DAY

The length of a day is relative to whether we measure by the Sun or other stars.

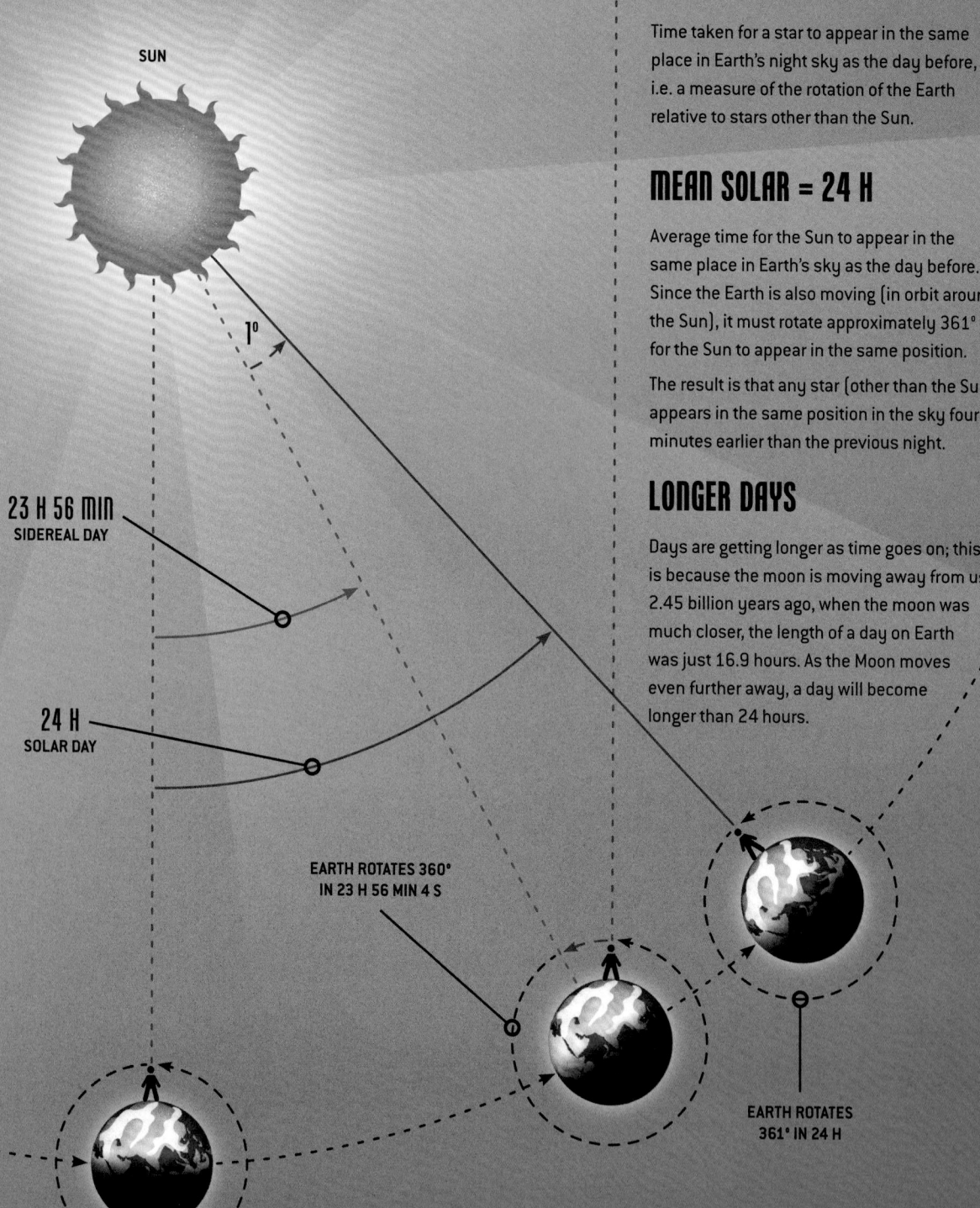

DISTANT STARS

SUN

1°

23 H 56 MIN
SIDEREAL DAY

24 H
SOLAR DAY

EARTH ROTATES 360°
IN 23 H 56 MIN 4 S

EARTH ROTATES
361° IN 24 H

SIDEREAL DAY = 23 H 56 MIN 4 S

Time taken for a star to appear in the same place in Earth's night sky as the day before, i.e. a measure of the rotation of the Earth relative to stars other than the Sun.

MEAN SOLAR = 24 H

Average time for the Sun to appear in the same place in Earth's sky as the day before. Since the Earth is also moving (in orbit around the Sun), it must rotate approximately 361° for the Sun to appear in the same position.

The result is that any star (other than the Sun) appears in the same position in the sky four minutes earlier than the previous night.

LONGER DAYS

Days are getting longer as time goes on; this is because the moon is moving away from us. 2.45 billion years ago, when the moon was much closer, the length of a day on Earth was just 16.9 hours. As the Moon moves even further away, a day will become longer than 24 hours.

THE RIGHT TIME

Einstein's theories of relativity show us that time is affected by gravity and by how fast you're moving. This has an impact on objects we launch into space.

TIME DILATION DUE TO GRAVITY

General relativity shows that the further you are from a massive body such as Earth, the faster time passes. See **NEPTUNE > GRAVITY AND THE DISCOVERY OF NEPTUNE** 154

  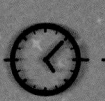

TIME DILATION DUE TO VELOCITY

Special relativity shows that time passes more slowly for fast-moving objects.

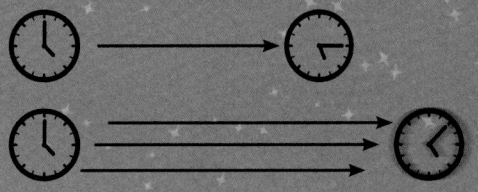

GLOBAL POSITIONING SATELLITE (GPS)

ALTITUDE: 20,000 km **SPEED:** 14,000 km/h

An object on Earth's surface moves at approximately 1,675 km/h as the Earth spins, and is lower in the gravitational field than a GPS satellite. This means that for the satellite time runs both slower due to its speed and faster as it is higher in the gravitational field. The gravitational dilation is more powerful so the GPS clock runs faster than one on Earth.

TIME RELATIVE TO EARTH CLOCK: Faster, by 38 millionths of a second per day

The effect might sound small, but GPS systems must compensate for the difference if satellite navigation and other technologies are to work properly. So whenever you use SatNav, you're travelling in time!

EARTH

INTERNATIONAL SPACE STATION

ALTITUDE: 408–410 km **SPEED:** 27,600 km/h

In low-Earth orbit and moving fast, so clocks onboard are slowed down more than they are sped up.

TIME RELATIVE TO EARTH CLOCK: Slower, by 28 millionths of a second per day.

07. MARS

MANTLE

* Thought to be similar to Earth's

NORTHERN ICE CAP

* Several km thick
* Largely water ice, which grows and shrinks with Martian seasons

NORTHERN CRUST

* Permafrost at high latitudes
* Northern plains with some cratering

CRUST

* Varies from 30 to 100 km thick — about the same as the Earth's continental crust
* Basalt, more like Earth's oceanic crust
* No tectonic plates

CORE

* Partly liquid core
* Diameter 3,400 km
* Evidence suggests that until ~4 billion years ago, this generated a magnetic field

ATMOSPHERE

* Mostly CO_2 with trace amounts of water

SOUTHERN CRUST

* Older, heavily cratered southern highlands
* About 4.5 billion years-old

SOUTHERN ICE CAP

* Ice consisting of water and CO_2, plus possible underground saltwater lake

EQUATOR

* Average temperature at equator −58ºC but can reach 0ºC at noon
* Equatorial frozen ocean, the same size as the North Sea on Earth

KEY FACTS

Diameter: 6,794 km
Average distance from Sun: 228 million km
Average distance from Earth: 78 million km
Day: 24 hours, 37 minutes
Year: 687 Earth days
Moons: 2

WOBBLY OBLIQUITY

Mars is now tilted 25.19° to the plane of its path around the Sun, but evidence suggests that this angle has varied between 5° and 35° over the past 125,000 years. This is thought to be due to Mars having no large moon to keep it stable on its axis.

SCALE

EARTH

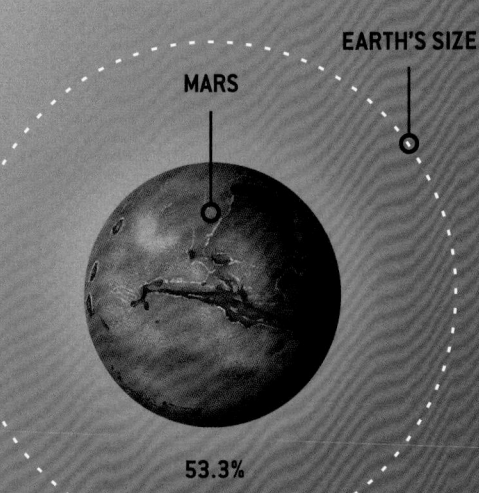

MARS

EARTH'S SIZE

**53.3%
OF EARTH**

NOW

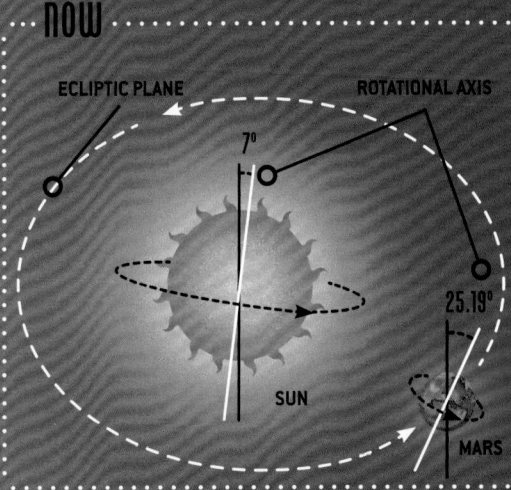

ECLIPTIC PLANE

ROTATIONAL AXIS

7°

25.19°

SUN

MARS

5°

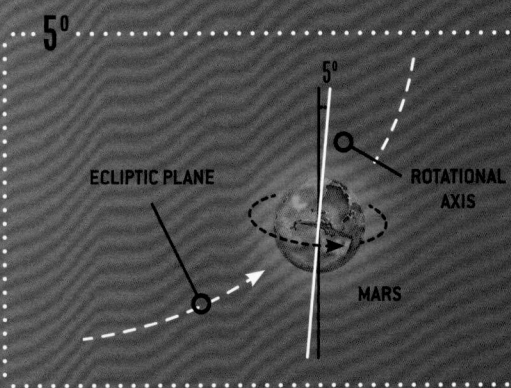

5°

ECLIPTIC PLANE

ROTATIONAL AXIS

MARS

35°

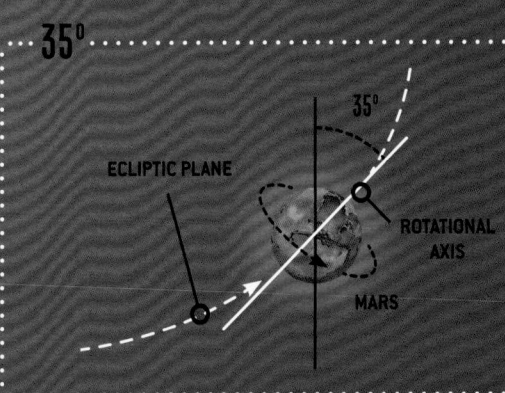

35°

ECLIPTIC PLANE

ROTATIONAL AXIS

MARS

SPACE JARGON – II

The technical language used by astronomers – made easy.

SUPERIOR PLANET

Mars, Jupiter, Saturn, Uranus and Neptune, which are further from
the Sun than the Earth and have larger orbital paths.

To read about Inferior Planets, see **MERCURY > SPACE JARGON – I**

SUN

MARS

EARTH

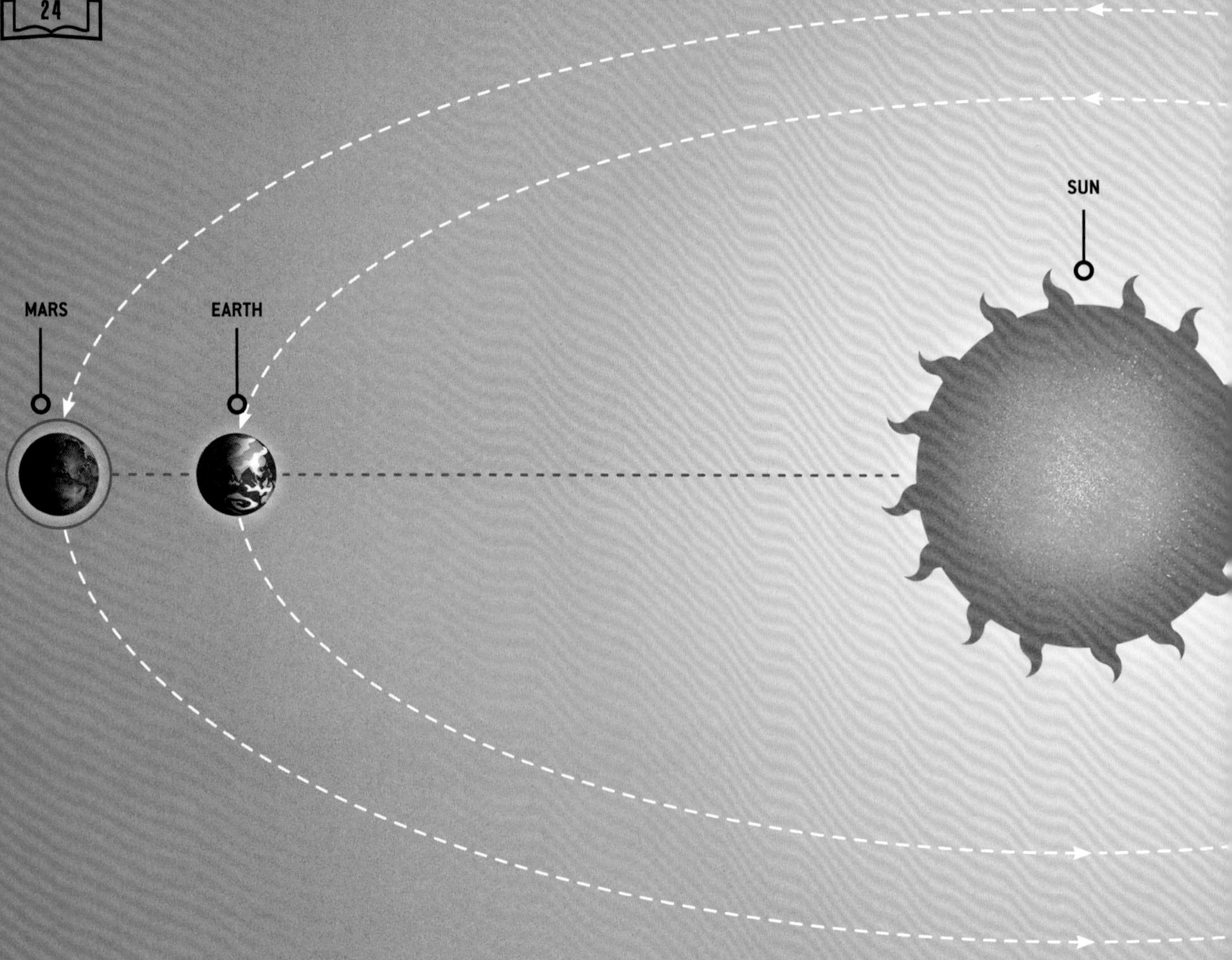

OPPOSITION

When Earth is directly between
the Sun and a superior planet

CONJUNCTION

When Mars or other superior planet is
on the far side of the Sun from Earth

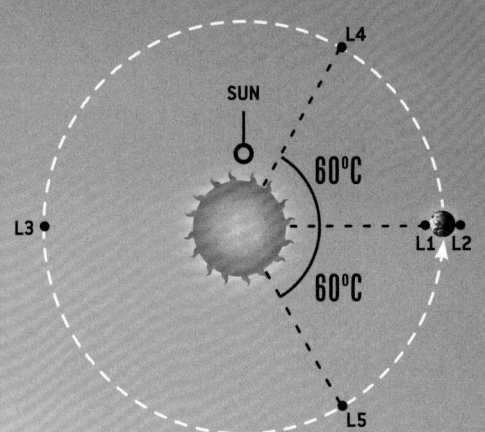

LAGRANGE POINTS

Lagrange points are locations in space where the gravitational forces of two larger bodies, such as the Sun and Earth, balance the forces felt by a smaller body, such as a satellite, creating an equilibrium where the smaller body can effectively be parked relative to the larger bodies.

L1, L2 AND L3 ARE UNSTABLE but a spacecraft needs less fuel to remain in fixed position. James Webb Space Telescope located at L2 since 24 January 2022.

IN THE SUN-EARTH SYSTEM:

✳ The Solar and Heliospheric Observatory Satellite (SOHO) is at **L1**

✳ **L2** is a good position for space telescopes – close enough to Earth for easy communication but with a good, unimpaired view of the cosmos

L4 AND L5 ARE STABLE but a spacecraft (and other objects) can remain in position without using fuel.

✳ Rocks called Trojans sit at the **L4** and / or **L5** points in the orbits of Venus, Earth, Mars, Jupiter, Neptune and Uranus.

See **JUPITER > TROJANS** 128

NOT TO SCALE

15 JULY 2067
MERCURY WILL OCCULT NEPTUNE

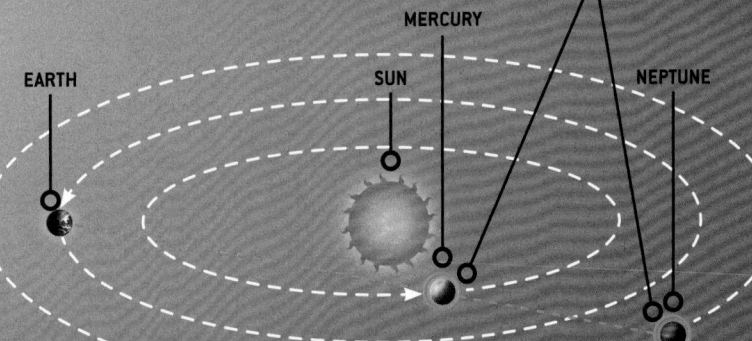

OCCULTATION

When one celestial body passes in front of another and so hides it, as seen from Earth

MISSIONS TO MARS

The probes and rovers we've landed — and crashed — on Mars.

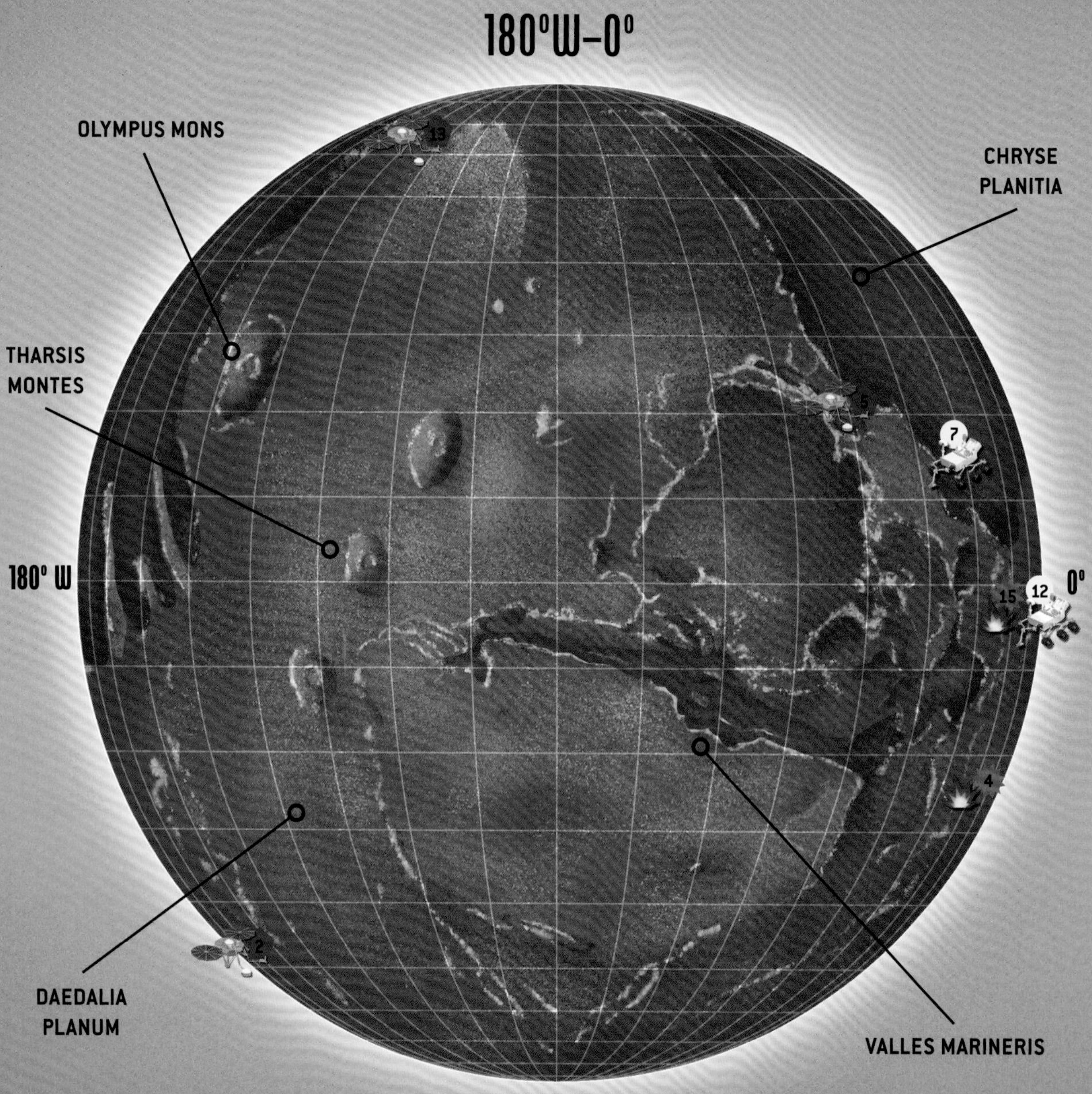

180ºW–0º

OLYMPUS MONS

CHRYSE PLANITIA

THARSIS MONTES

180º W

0º

13

?

15
12
1

4

2

DAEDALIA PLANUM

VALLES MARINERIS

KEY TO LANDINGS

Rover

Probe

Crash

1. **Mars 2:** 27 November 1971
2. **Mars 3:** 3 December 1971
3. **Mars 5:** 28 February 1974
4. **Mars 6:** 12 March 1974
5. **Viking 1:** 20 July 1976
6. **Viking 2:** 7 August 1976

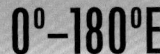

0°–180°E

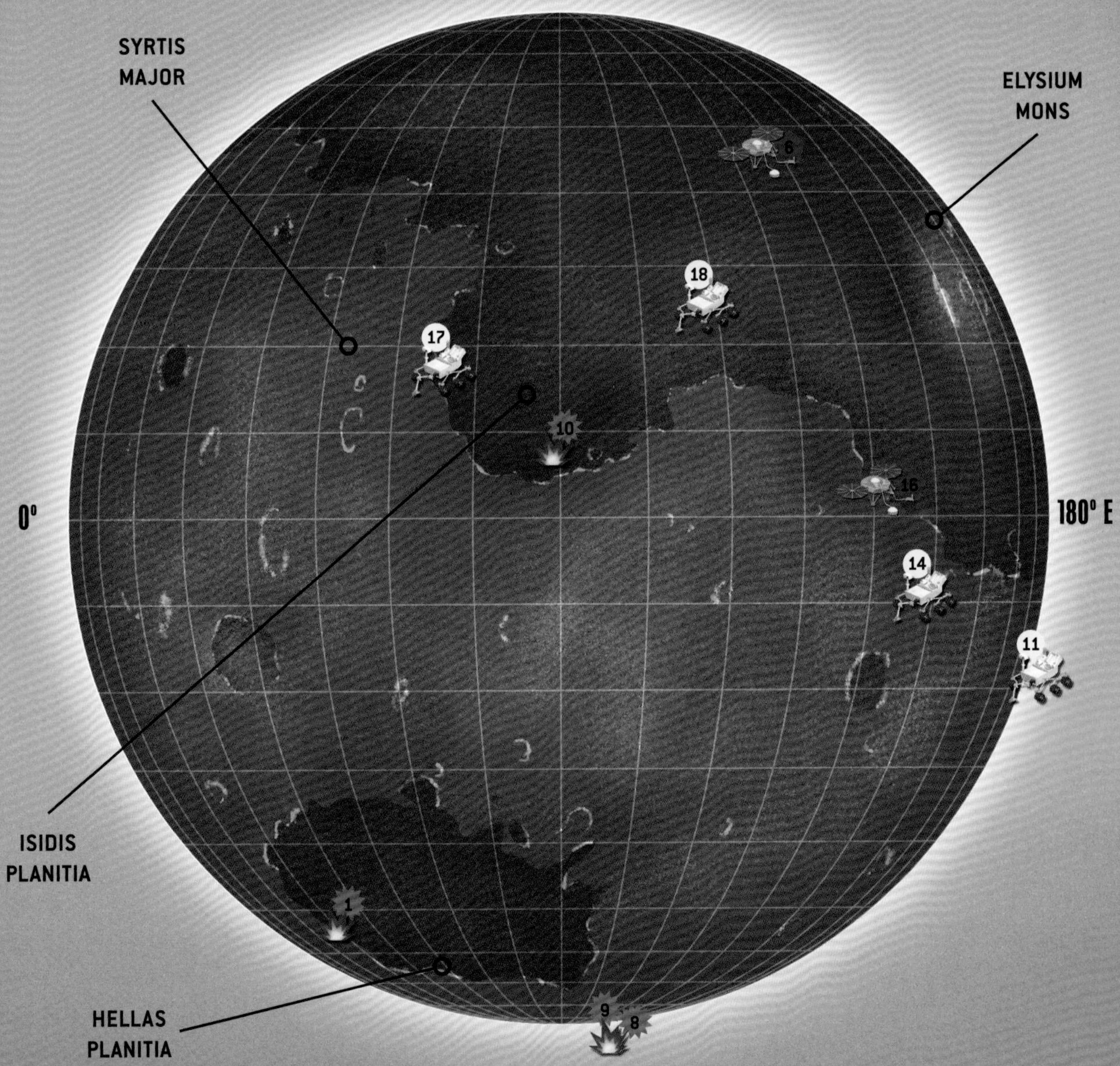

SYRTIS MAJOR
ELYSIUM MONS
0°
180° E
ISIDIS PLANITIA
HELLAS PLANITIA
6
18
17
10
16
14
11
1
9
8

- **7. Pathfinder and Sojourner:** 4 July 1997
- **8. Polar Lander:** 3 December 1999
- **9. Deep Space 2:** 3 December 1999
- **10. Beagle 2:** 25 December 2003
- **11. Spirit:** 4 January 2004
- **12. Opportunity:** 25 January 2004
- **13. Phoenix:** 25 May 2008
- **14. Curiosity:** 6 August 2012
- **15. Schiaparelli:** 19 October 2016
- **16. InSight:** 26 November 2018
- **17. Perseverance rover, Ingenuity helicopter:** 18 February 2021
- **18. Tianwen-1 lander, Zhurong rover:** 14 May 2021
- **19. Mariner 9:** Burned up in atmosphere or crashed on surface 2022–23, site unknown

SOJOURNER, SPIRIT AND CURIOSITY

The incredible journeys of the Mars Explorer Rovers.

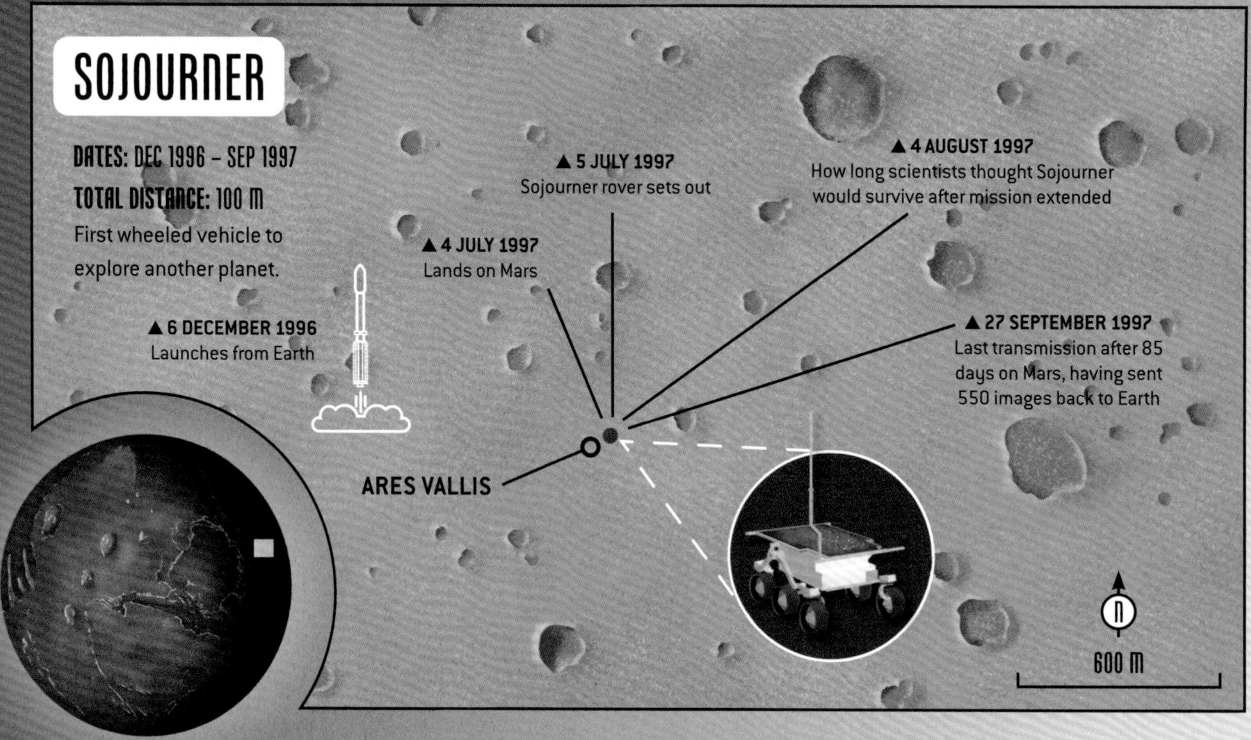

SOJOURNER

DATES: DEC 1996 – SEP 1997

TOTAL DISTANCE: 100 M

First wheeled vehicle to explore another planet.

▲ **6 DECEMBER 1996**
Launches from Earth

▲ **4 JULY 1997**
Lands on Mars

▲ **5 JULY 1997**
Sojourner rover sets out

▲ **4 AUGUST 1997**
How long scientists thought Sojourner would survive after mission extended

▲ **27 SEPTEMBER 1997**
Last transmission after 85 days on Mars, having sent 550 images back to Earth

ARES VALLIS

N

600 M

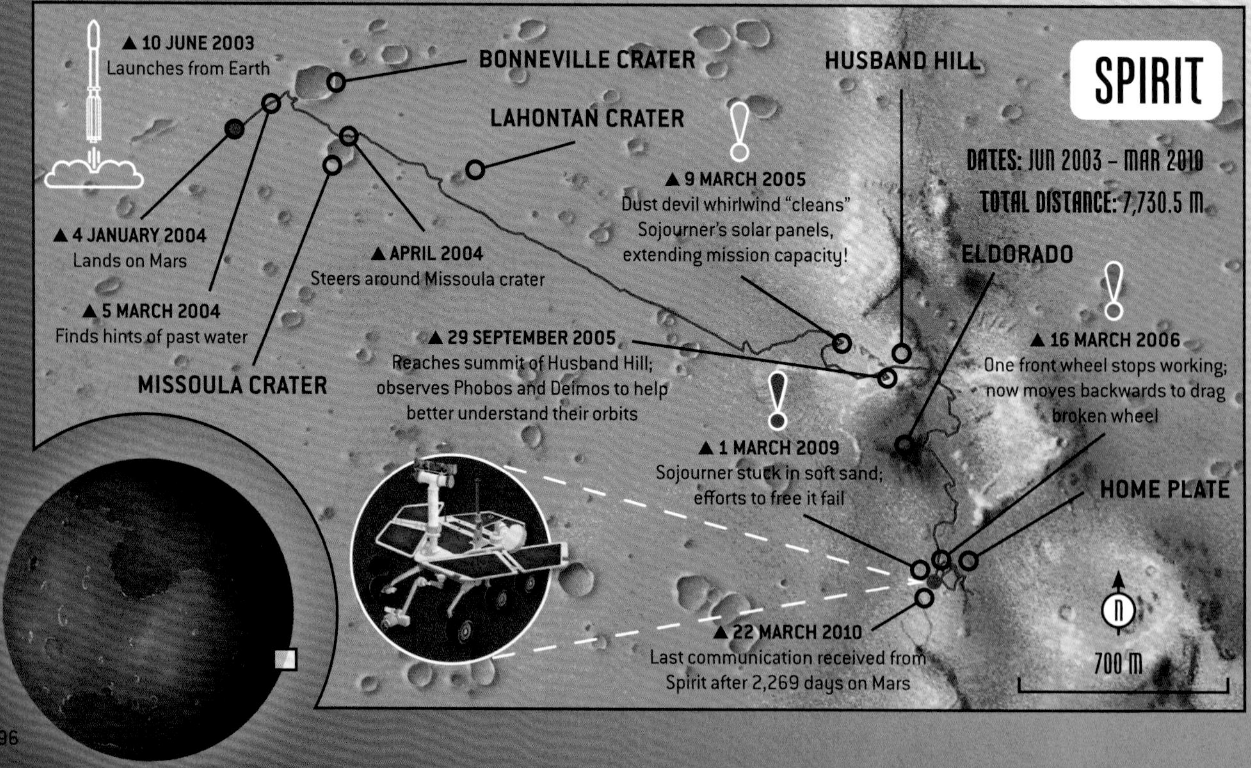

▲ **10 JUNE 2003**
Launches from Earth

BONNEVILLE CRATER

LAHONTAN CRATER

HUSBAND HILL

SPIRIT

DATES: JUN 2003 – MAR 2010

TOTAL DISTANCE: 7,730.5 M

▲ **4 JANUARY 2004**
Lands on Mars

▲ **5 MARCH 2004**
Finds hints of past water

▲ **APRIL 2004**
Steers around Missoula crater

▲ **9 MARCH 2005**
Dust devil whirlwind "cleans" Sojourner's solar panels, extending mission capacity!

ELDORADO

▲ **16 MARCH 2006**
One front wheel stops working; now moves backwards to drag broken wheel

▲ **29 SEPTEMBER 2005**
Reaches summit of Husband Hill; observes Phobos and Deimos to help better understand their orbits

MISSOULA CRATER

▲ **1 MARCH 2009**
Sojourner stuck in soft sand; efforts to free it fail

HOME PLATE

▲ **22 MARCH 2010**
Last communication received from Spirit after 2,269 days on Mars

N

700 M

CURIOSITY

DATES: NOV 2011 – ONGOING*

TOTAL DISTANCE: 31,130 M (TO DEC 2023)

* As of publication date

2 KM

N

▲ **26 NOVEMBER 2011**
Launches from Earth

BRADBURY LANDING

YELLOWKNIFE BAY

▲ **6 AUGUST 2012**
Lands on Mars

▲ **29 AUGUST 2012**
After days of checks, Curiosity
begins roaming

▲ **4 FEBRUARY 2013**
First use of onboard
drill to mine samples

▲ **16 JULY 2013**
Has travelled 1 km since landing. Rover
has transmitted 190 gigabits of data
to Earth in that time, including 70,000
images, while it has fired laser 75,000
times at 2,000 targets.

▲ **27 SEPTEMBER 2012**
Finds evidence of
ancient stream bed

▲ **3 JUNE 2014**
Observes Mercury transiting
the Sun, the first transit seen
from another planet

▲ **9 DECEMBER 2013**
NASA publishes six articles
in Science full of discoveries
made by Curiosity

▲ **24 JUNE 2014**
Completes Martian year of
687 Earth days on Mars

▲ **11 SEPTEMBER 2014**
After total of 6.9 km travelled,
Curiosity reaches slope of Aeolis
Mons — where it will learn about
the history of Mars

▲ **16 DECEMBER 2014**
Detects 10x localised spike
in methane in atmosphere

▲ **14 MARCH 2015**
Detects nitrogen produced by
heating Martian sediments

▲ **13 DECEMBER 2016**
Curiosity has travelled
15 km and climbed 165 m

▲ **4 JUNE 2018**
Problems with drill solved

▲ **3 NOVEMBER 2023**
Curiosity has now been on Mars for
4,000 'sols', or Martian solar days
(see page 99 for explanation)

▲ **3 OCTOBER 2018**
Back-up computer now
being used

▲ **17 & 27 MARCH 2019**
Observes solar eclipse by
Martian moons Deimos
and Phobos respectively

▲ **31 DECEMBER 2023**
Now on its fourth extended mission since
arriving on Mars, Curiosity continues to explore
Mount Sharp and conduct experiments…

ONGOING

OPPORTUNITY

The incredible journeys of the Mars Explorer Rovers.

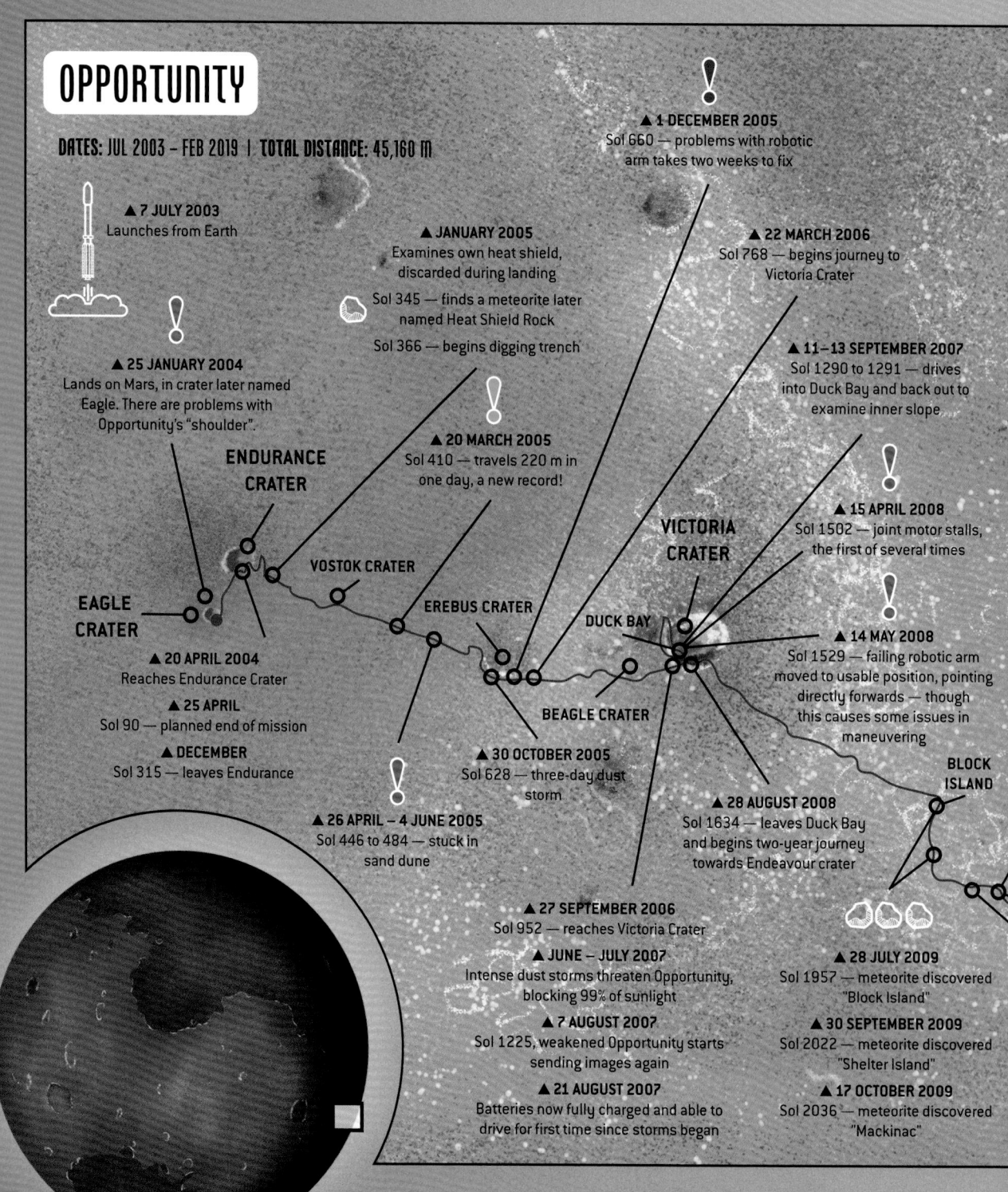

OPPORTUNITY

DATES: JUL 2003 – FEB 2019 | **TOTAL DISTANCE:** 45,160 M

▲ 7 JULY 2003
Launches from Earth

▲ 25 JANUARY 2004
Lands on Mars, in crater later named Eagle. There are problems with Opportunity's "shoulder".

ENDURANCE CRATER

▲ JANUARY 2005
Examines own heat shield, discarded during landing

Sol 345 — finds a meteorite later named Heat Shield Rock

Sol 366 — begins digging trench

▲ 20 MARCH 2005
Sol 410 — travels 220 m in one day, a new record!

VOSTOK CRATER

EAGLE CRATER

▲ 20 APRIL 2004
Reaches Endurance Crater

▲ 25 APRIL
Sol 90 — planned end of mission

▲ DECEMBER
Sol 315 — leaves Endurance

▲ 26 APRIL – 4 JUNE 2005
Sol 446 to 484 — stuck in sand dune

EREBUS CRATER

▲ 30 OCTOBER 2005
Sol 628 — three-day dust storm

BEAGLE CRATER

▲ 1 DECEMBER 2005
Sol 660 — problems with robotic arm takes two weeks to fix

▲ 22 MARCH 2006
Sol 768 — begins journey to Victoria Crater

▲ 11–13 SEPTEMBER 2007
Sol 1290 to 1291 — drives into Duck Bay and back out to examine inner slope

VICTORIA CRATER

DUCK BAY

▲ 15 APRIL 2008
Sol 1502 — joint motor stalls, the first of several times

▲ 14 MAY 2008
Sol 1529 — failing robotic arm moved to usable position, pointing directly forwards — though this causes some issues in maneuvering

BLOCK ISLAND

▲ 28 AUGUST 2008
Sol 1634 — leaves Duck Bay and begins two-year journey towards Endeavour crater

▲ 27 SEPTEMBER 2006
Sol 952 — reaches Victoria Crater

▲ JUNE – JULY 2007
Intense dust storms threaten Opportunity, blocking 99% of sunlight

▲ 7 AUGUST 2007
Sol 1225, weakened Opportunity starts sending images again

▲ 21 AUGUST 2007
Batteries now fully charged and able to drive for first time since storms began

▲ 28 JULY 2009
Sol 1957 — meteorite discovered "Block Island"

▲ 30 SEPTEMBER 2009
Sol 2022 — meteorite discovered "Shelter Island"

▲ 17 OCTOBER 2009
Sol 2036 — meteorite discovered "Mackinac"

Missions are usually measured in "sols" or Martian days of 24 hours, 39 minutes and 35.2 seconds — or 1.027 Earth days.

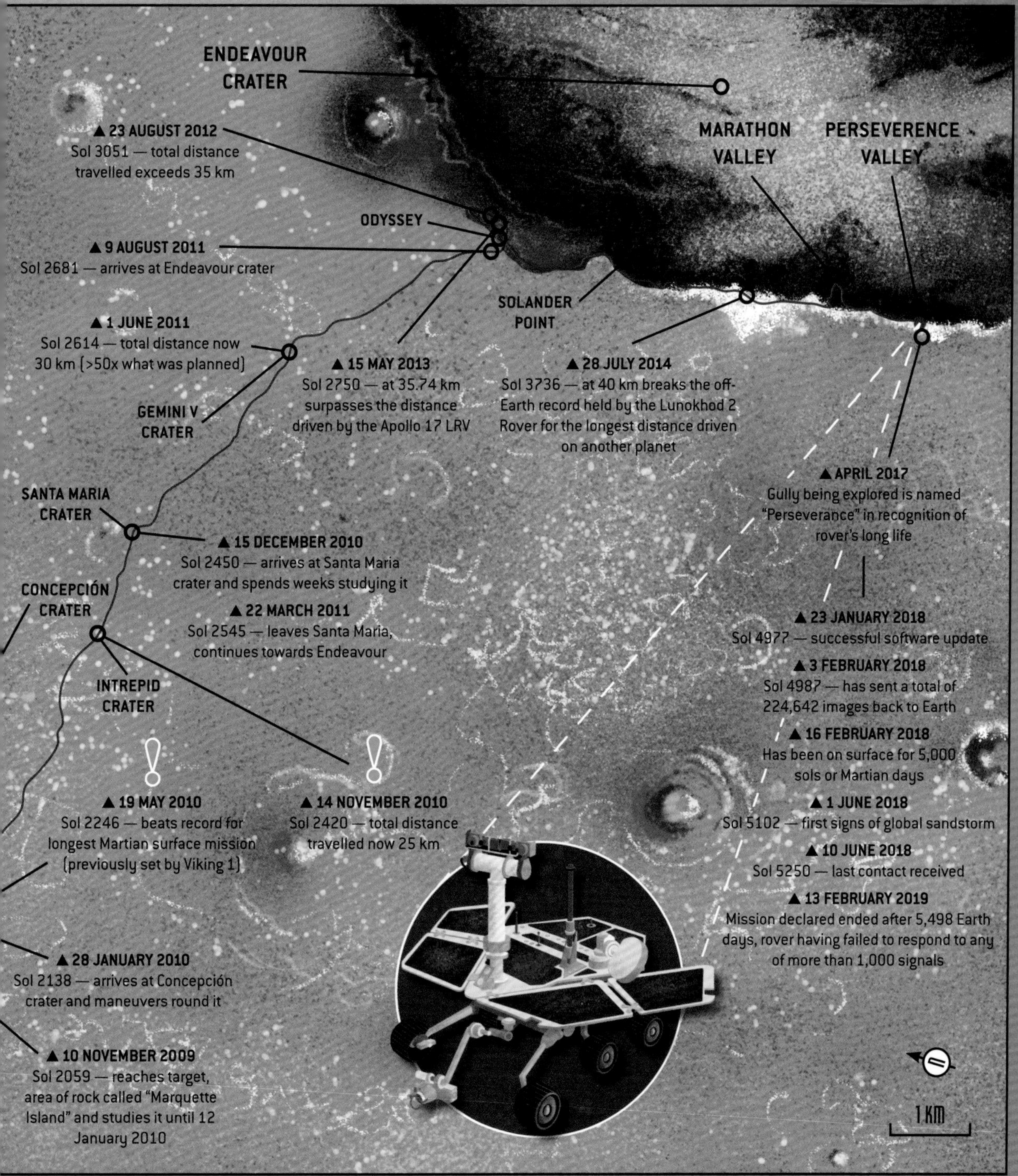

ENDEAVOUR CRATER

MARATHON VALLEY

PERSEVERENCE VALLEY

▲ **23 AUGUST 2012**
Sol 3051 — total distance travelled exceeds 35 km

ODYSSEY

▲ **9 AUGUST 2011**
Sol 2681 — arrives at Endeavour crater

SOLANDER POINT

▲ **1 JUNE 2011**
Sol 2614 — total distance now 30 km (>50x what was planned)

▲ **15 MAY 2013**
Sol 2750 — at 35.74 km surpasses the distance driven by the Apollo 17 LRV

▲ **28 JULY 2014**
Sol 3736 — at 40 km breaks the off-Earth record held by the Lunokhod 2 Rover for the longest distance driven on another planet

GEMINI V CRATER

▲ **APRIL 2017**
Gully being explored is named "Perseverance" in recognition of rover's long life

SANTA MARIA CRATER

▲ **15 DECEMBER 2010**
Sol 2450 — arrives at Santa Maria crater and spends weeks studying it

▲ **23 JANUARY 2018**
Sol 4977 — successful software update

CONCEPCIÓN CRATER

▲ **22 MARCH 2011**
Sol 2545 — leaves Santa Maria, continues towards Endeavour

▲ **3 FEBRUARY 2018**
Sol 4987 — has sent a total of 224,642 images back to Earth

INTREPID CRATER

▲ **16 FEBRUARY 2018**
Has been on surface for 5,000 sols or Martian days

▲ **19 MAY 2010**
Sol 2246 — beats record for longest Martian surface mission (previously set by Viking 1)

▲ **14 NOVEMBER 2010**
Sol 2420 — total distance travelled now 25 km

▲ **1 JUNE 2018**
Sol 5102 — first signs of global sandstorm

▲ **10 JUNE 2018**
Sol 5250 — last contact received

▲ **13 FEBRUARY 2019**
Mission declared ended after 5,498 Earth days, rover having failed to respond to any of more than 1,000 signals

▲ **28 JANUARY 2010**
Sol 2138 — arrives at Concepción crater and maneuvers round it

▲ **10 NOVEMBER 2009**
Sol 2059 — reaches target, area of rock called "Marquette Island" and studies it until 12 January 2010

1 KM

PERSEVERANCE, INGENUITY AND ZHURONG

The incredible journeys of the Mars Explorer Rovers.

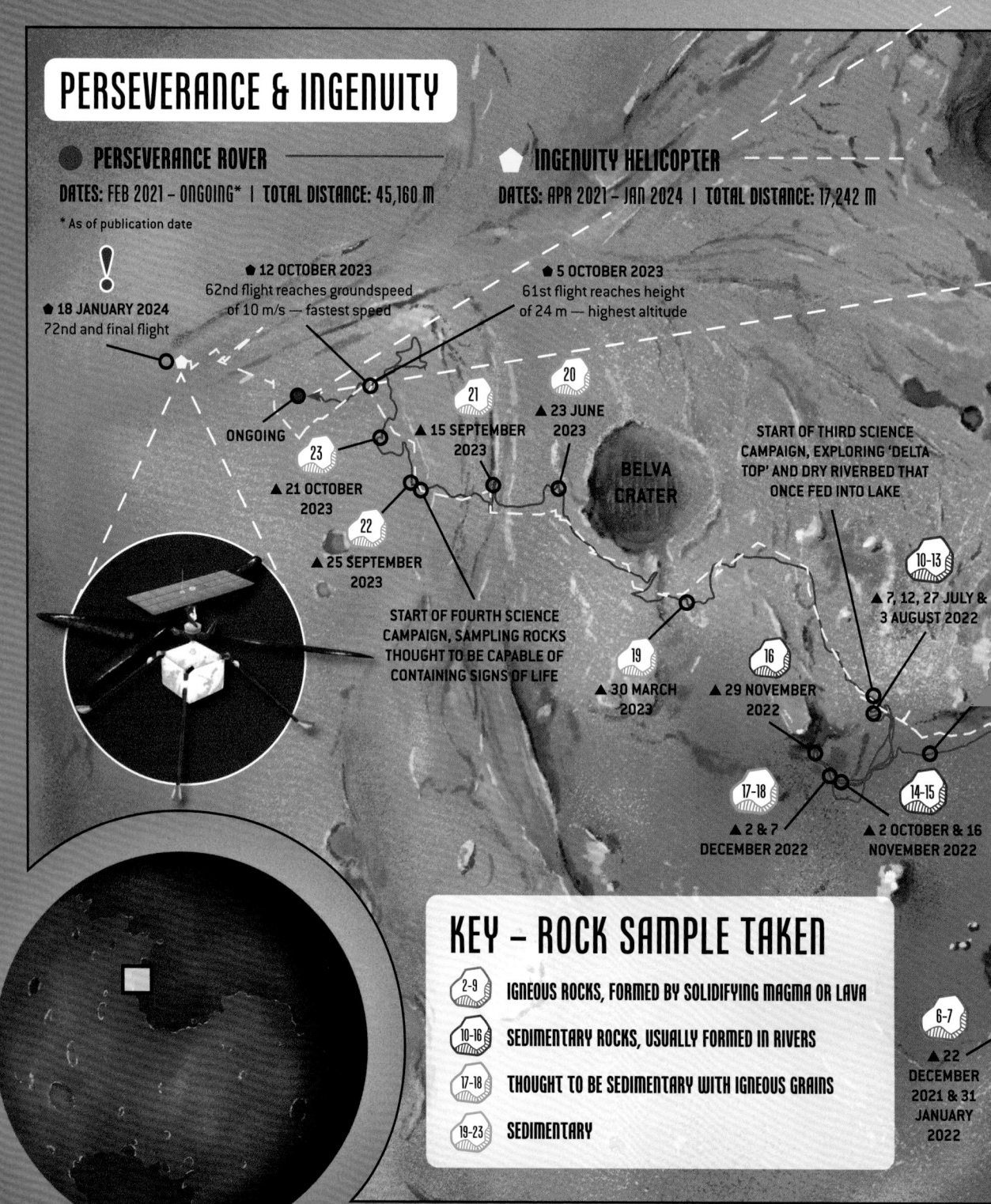

PERSEVERANCE & INGENUITY

● **PERSEVERANCE ROVER**
DATES: FEB 2021 – ONGOING* | TOTAL DISTANCE: 45,160 M

⬠ **INGENUITY HELICOPTER**
DATES: APR 2021 – JAN 2024 | TOTAL DISTANCE: 17,242 M

* As of publication date

⬠ **18 JANUARY 2024**
72nd and final flight

⬠ **12 OCTOBER 2023**
62nd flight reaches groundspeed
of 10 m/s — fastest speed

⬠ **5 OCTOBER 2023**
61st flight reaches height
of 24 m — highest altitude

ONGOING

21
▲ 15 SEPTEMBER
2023

20
▲ 23 JUNE
2023

23
▲ 21 OCTOBER
2023

22
▲ 25 SEPTEMBER
2023

**BELVA
CRATER**

START OF THIRD SCIENCE
CAMPAIGN, EXPLORING 'DELTA
TOP' AND DRY RIVERBED THAT
ONCE FED INTO LAKE

10-13
▲ 7, 12, 27 JULY &
3 AUGUST 2022

START OF FOURTH SCIENCE
CAMPAIGN, SAMPLING ROCKS
THOUGHT TO BE CAPABLE OF
CONTAINING SIGNS OF LIFE

19
▲ 30 MARCH
2023

16
▲ 29 NOVEMBER
2022

17-18
▲ 2 & 7
DECEMBER 2022

14-15
▲ 2 OCTOBER & 16
NOVEMBER 2022

6-7
▲ 22
DECEMBER
2021 & 31
JANUARY
2022

KEY – ROCK SAMPLE TAKEN

2-9 IGNEOUS ROCKS, FORMED BY SOLIDIFYING MAGMA OR LAVA

10-16 SEDIMENTARY ROCKS, USUALLY FORMED IN RIVERS

17-18 THOUGHT TO BE SEDIMENTARY WITH IGNEOUS GRAINS

19-23 SEDIMENTARY

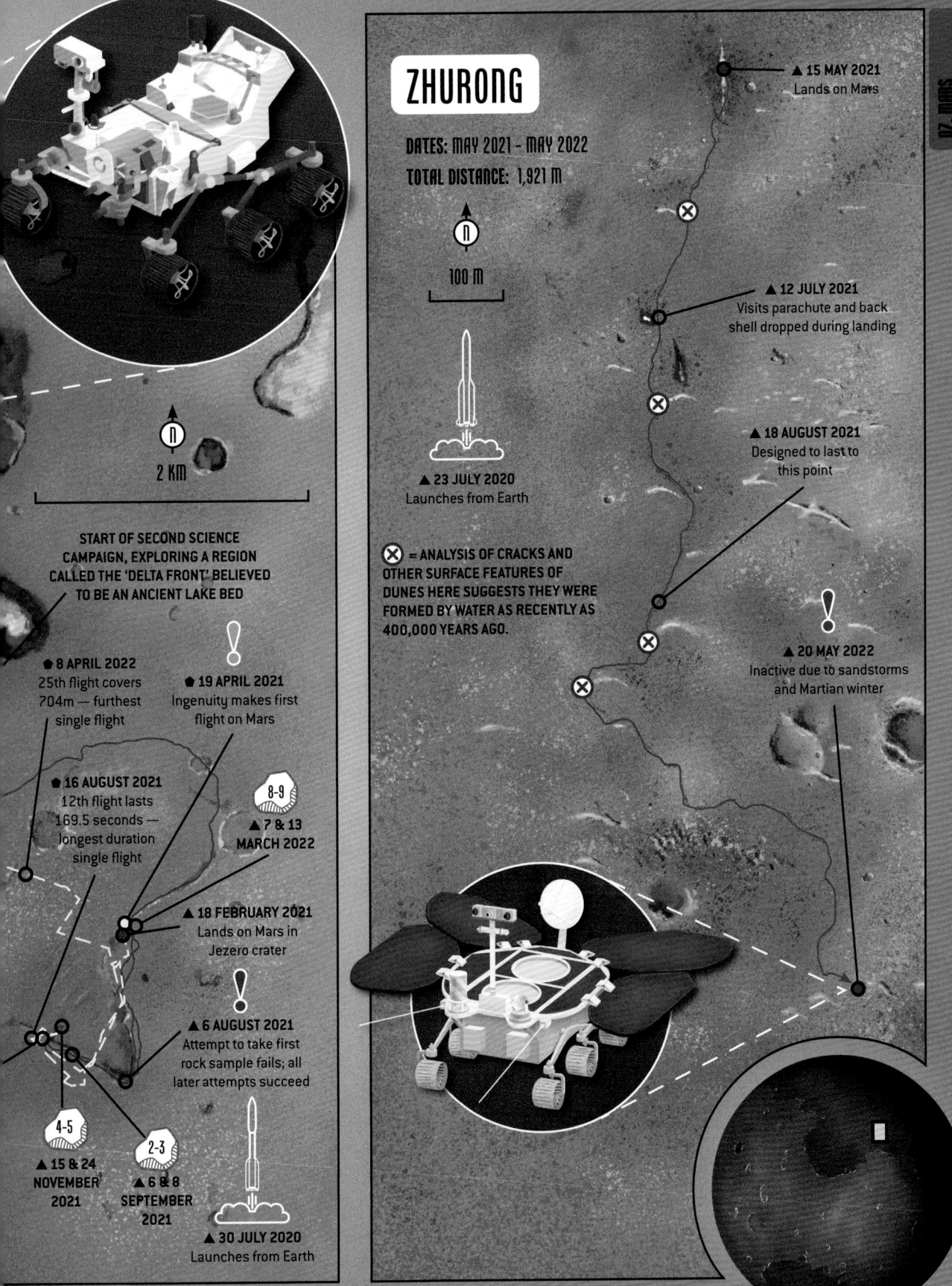

ZHURONG

DATES: MAY 2021 – MAY 2022
TOTAL DISTANCE: 1,921 M

N

100 M

▲ 23 JULY 2020
Launches from Earth

⊗ = ANALYSIS OF CRACKS AND OTHER SURFACE FEATURES OF DUNES HERE SUGGESTS THEY WERE FORMED BY WATER AS RECENTLY AS 400,000 YEARS AGO.

▲ 15 MAY 2021
Lands on Mars

▲ 12 JULY 2021
Visits parachute and back shell dropped during landing

▲ 18 AUGUST 2021
Designed to last to this point

▲ 20 MAY 2022
Inactive due to sandstorms and Martian winter

N

2 KM

START OF SECOND SCIENCE CAMPAIGN, EXPLORING A REGION CALLED THE 'DELTA FRONT' BELIEVED TO BE AN ANCIENT LAKE BED

● 8 APRIL 2022
25th flight covers 704m — furthest single flight

● 19 APRIL 2021
Ingenuity makes first flight on Mars

● 16 AUGUST 2021
12th flight lasts 169.5 seconds — longest duration single flight

8-9
▲ 7 & 13
MARCH 2022

▲ 18 FEBRUARY 2021
Lands on Mars in Jezero crater

▲ 6 AUGUST 2021
Attempt to take first rock sample fails; all later attempts succeed

4-5
▲ 15 & 24
NOVEMBER
2021

2-3
▲ 6 & 8
SEPTEMBER
2021

▲ 30 JULY 2020
Launches from Earth

AIN'T NO MOUNTAIN HIGH ENOUGH

Volcanoes are among the tallest mountains in the Solar System.

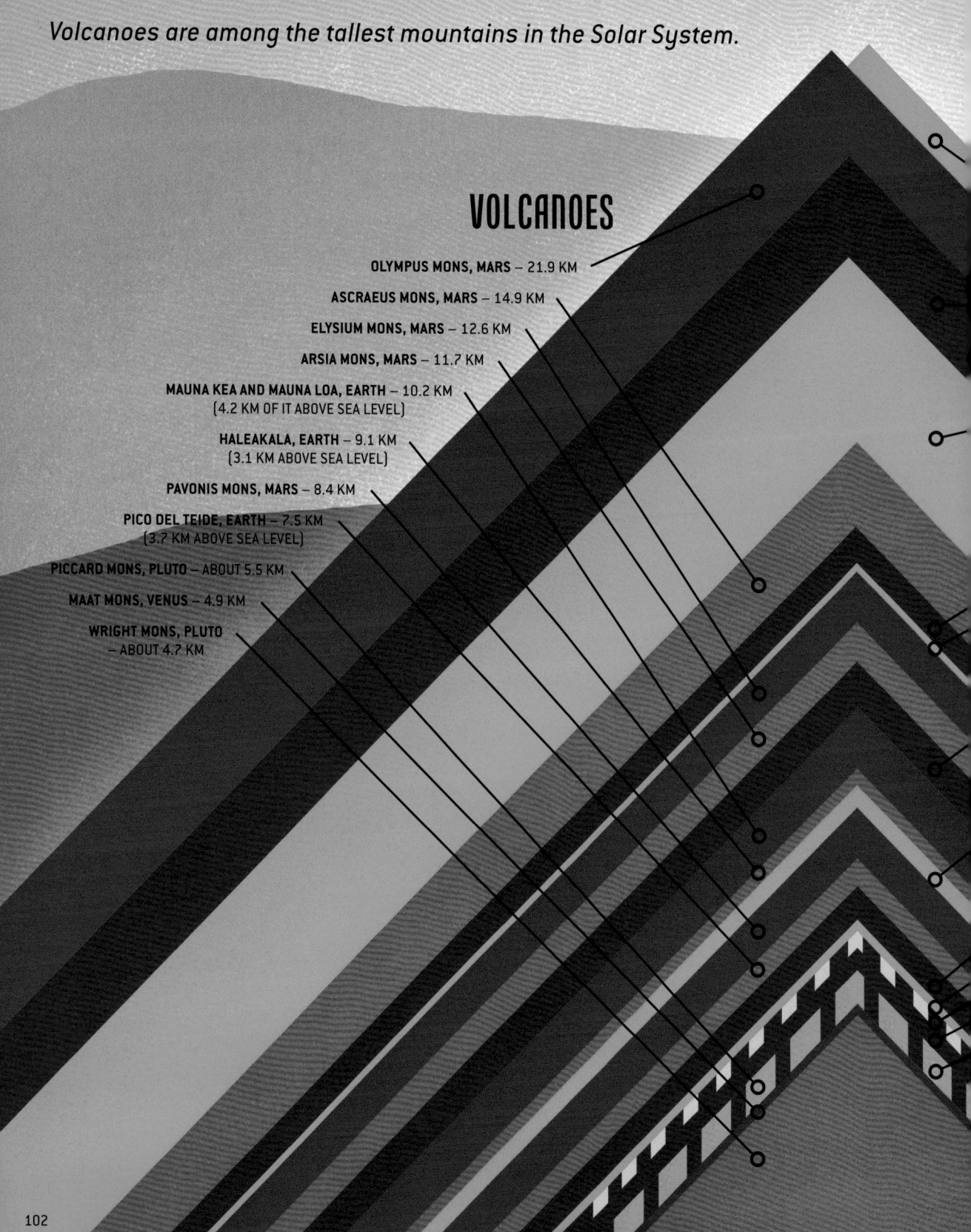

VOLCANOES

OLYMPUS MONS, MARS – 21.9 KM

ASCRAEUS MONS, MARS – 14.9 KM

ELYSIUM MONS, MARS – 12.6 KM

ARSIA MONS, MARS – 11.7 KM

MAUNA KEA AND MAUNA LOA, EARTH – 10.2 KM
(4.2 KM OF IT ABOVE SEA LEVEL)

HALEAKALA, EARTH – 9.1 KM
(3.1 KM ABOVE SEA LEVEL)

PAVONIS MONS, MARS – 8.4 KM

PICO DEL TEIDE, EARTH – 7.5 KM
(3.7 KM ABOVE SEA LEVEL)

PICCARD MONS, PLUTO – ABOUT 5.5 KM

MAAT MONS, VENUS – 4.9 KM

WRIGHT MONS, PLUTO
– ABOUT 4.7 KM

KEY

 VOLCANO

 CRYOVOLCANO EJECTING WATER / ICE

EJECTS NITROGEN, METHANE, DUST

ACTIVE VOLCANOES

 EARTH

 IO — most volcanically active body in Solar System

 ENCELADUS, moon of Saturn

 TITAN, moon of Saturn

 TRITON, moon of Neptune

 PLUTO — possibly

OTHER KINDS OF MOUNTAIN

CENTRAL PEAK OF RHEASILVIA CRATER, ASTEROID VESTA — 22 KM

EQUATORIAL RIDGE, IAPETUS (MOON OF SATURN) — 20 KM

SOUTH BOÖSAULE, IO (MOON OF JUPITER) — 18.2 KM

EUBOEA MONTES, IO — 13.4 KM

EAST RIDGE OF IONIAN MONS, IO — 12.7 KM

UNNAMED MOUNTAIN ON OBERON, MOON OF URANUS — 11 KM

MOUNT EVEREST, EARTH — 8.8 KM (MEASURED FROM SEA LEVEL)

CENTRAL PEAK OF HERSCHEL CRATE, MIMAS (MOON OF SATURN) — 7 KM

SKADI MONS, VENUS — 6.4 KM

ANSERIS MONS, MARS — 6.2 KM

TENZING MONTES, PLUTO — ABOUT 6.2 KM

DENALI, EARTH — 5.9 KM

MONS HUGYENS, MOON — 5.5 KM

AELOIS MONS, MARS — 5.5 KM

PHOBOS AND DEIMOS

All about the moons of Mars.

PHOBOS

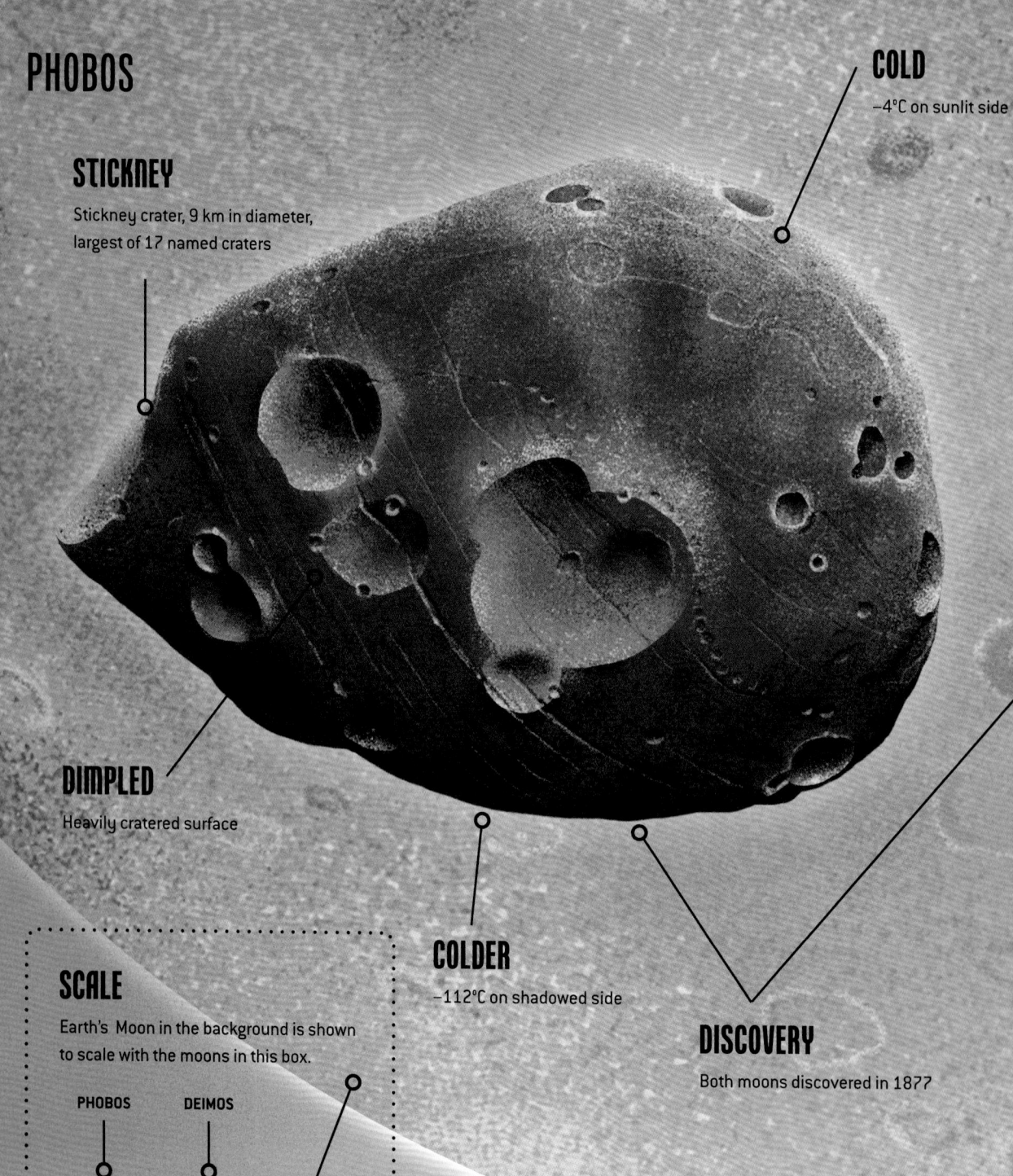

STICKNEY

Stickney crater, 9 km in diameter, largest of 17 named craters

COLD

−4°C on sunlit side

DIMPLED

Heavily cratered surface

COLDER

−112°C on shadowed side

DISCOVERY

Both moons discovered in 1877

SCALE

Earth's Moon in the background is shown to scale with the moons in this box.

PHOBOS   DEIMOS

MOON

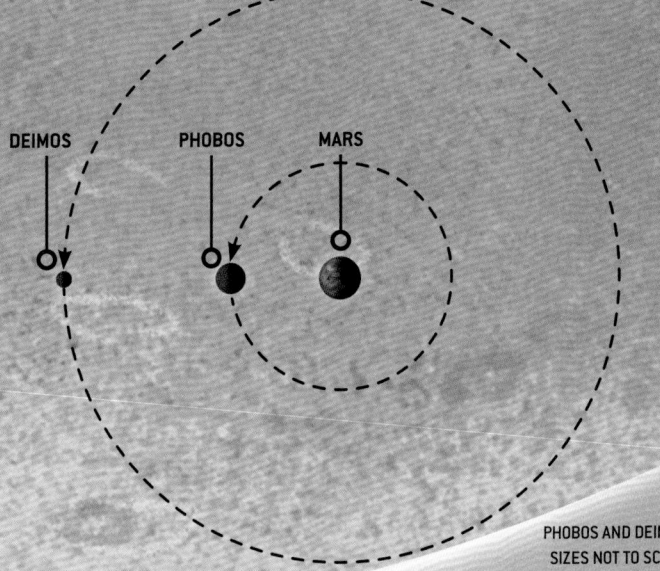

DEIMOS

SWIFT CRATER

VOLTAIRE CRATER

SMALLER

56% size of Phobos

SMOOTHIE

Much smoother than Phobos,
with only two named craters

ECLIPSES

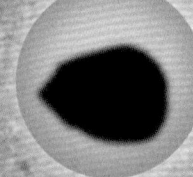

PHOBOS
Phobos eclipses the Sun,
seen from Martian surface

DEIMOS
Deimos transits the Sun,
seen from Martian surface

MOON
Moon eclipses Sun, seen from Earth.
See **MOON > ECLIPSES** 68

ORBITS

Phobos orbits Mars more closely than any other known planetary
moon — completing an orbit faster than Mars rotates! From the Martian
surface, it rises and sets twice each Martian day.

DEIMOS   PHOBOS   MARS

PHOBOS AND DEIMOS
SIZES NOT TO SCALE

105

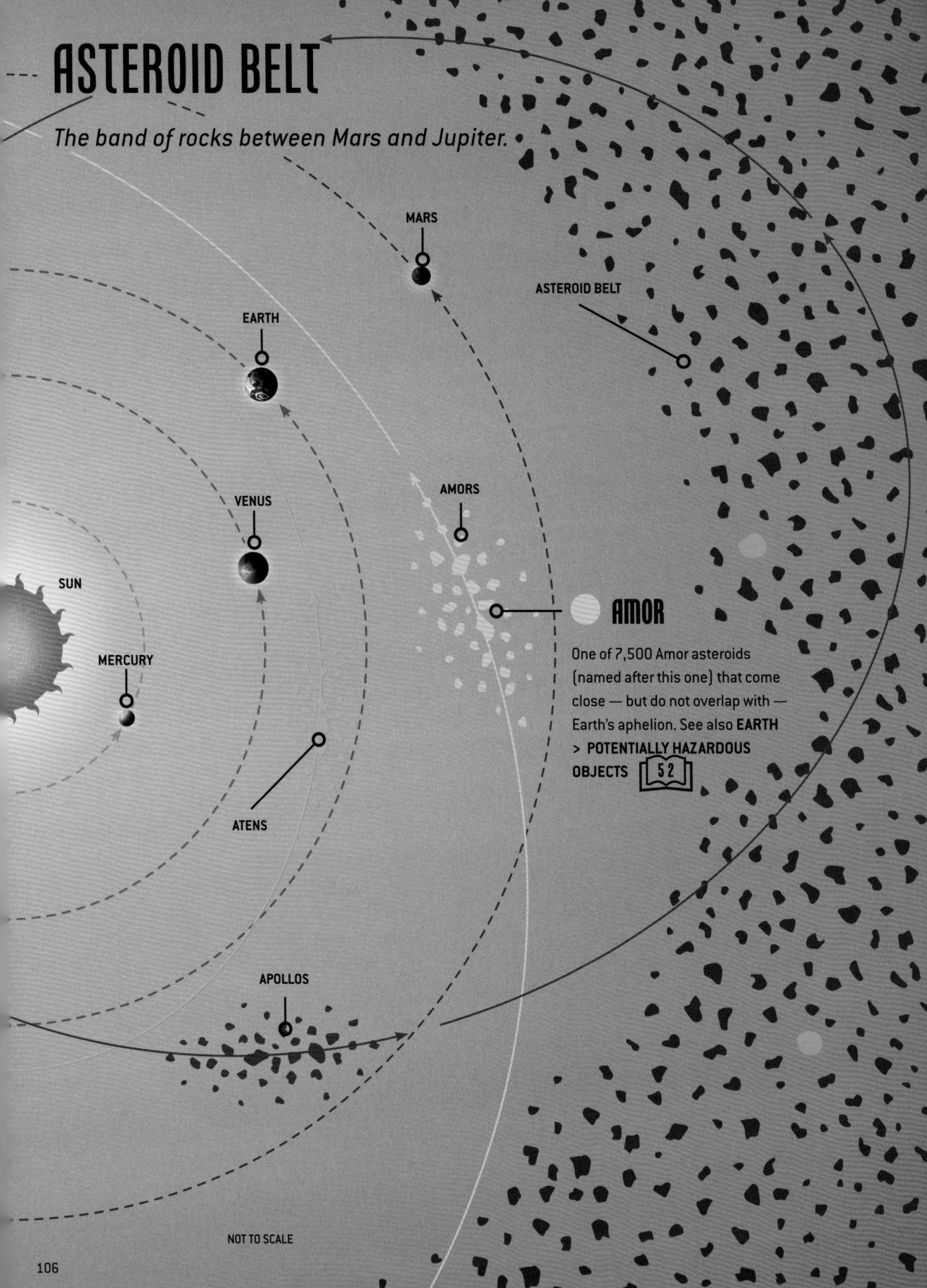

ASTEROID BELT

The band of rocks between Mars and Jupiter.

MARS

ASTEROID BELT

EARTH

AMORS

VENUS

AMOR

One of 7,500 Amor asteroids
(named after this one) that come
close — but do not overlap with —
Earth's aphelion. See also EARTH
> POTENTIALLY HAZARDOUS
OBJECTS 52

SUN

MERCURY

ATENS

APOLLOS

NOT TO SCALE

MAIN ASTEROID BELT

Millions of fragments of rock left over from the dawn of the Solar System but not enough to form a planet — the total mass is just 4% of Earth's Moon.

* >75% = "C-type" asteroids of dark, coal-black carbon.

* 17% = "S-type" bright silicates, especially in inner asteroid belt

CERES

GREEKS AND TROJANS

More than a million objects larger than 1 km share Jupiter's orbital path: "Greeks" ahead of the planet, "Trojans" behind it.
See JUPITER > TROJANS 128

GREEKS

HILDAS

BIG FOUR

The four largest objects in the asteroid belt = half its total mass:

* Dwarf planet **Ceres** (diameter ≅ 950 km, aphelion 3 AU, perihelion 2.6 AU)

* **Vesta** (mean diameter 525 km, distance from Sun 2.6–2.1 AU)

* **Pallas** (512 km, 3.4–.2.1 AU)

* **Hygiea** (434 km, 3.5–2.8 AU)

HILDA ASTEROIDS

More than 4,000 rocky objects, each with an elliptical orbit but together forming a stable triangle shape.

JUPITER

THE GREAT WALL OF SPACE?

Although there are lots of rocks in the asteroid belt, they're so spaced apart it's unlikely a space probe would hit one it wasn't specifically aiming for!

TROJANS

08. JUPITER

CLOUD TOPS

* Temperature at cloud tops = −130°C

BELTS

* Warmer, lower hydrocarbons — which produce the dark colour

LIQUID HYDROGEN

* At 1,000 km, deep ocean of molecular hydrogen where pressure high enough to liquefy atmosphere

UPPER ATMOSPHERE

* No solid surface — outermost layer of cloud largely hydrogen and helium
* 1,000 km thick

CORE

* We're unsure if Jupiter has a central core of solid material or a thick, dense and super-hot "soup"
* 32,000 km in diameter
* About 30,000°C

HALO RING

MAIN RING

AMALTHEA GOSSAMER RING

THEBE GOSSAMER RING

RINGS

Faint rings of dust, difficult to see from Earth:

* Halo ring, 30,500 km wide and 12,500 km thick
* Main ring, 6,500 km wide and between 30 and 300 km thick
* Amalthea gossamer ring, 53,000 km wide and 2,000 km thick
* Thebe gossamer ring, 97,000 km wide and 8,400 km thick

LIQUID METAL HYDROGEN

* At 50,000 km, liquid metal hydrogen because pressure is so high
* The rotation of this liquid metal generates magnetic field

ZONES

Brighter, cooler clouds of (in descending order):

* Ammonia ice crystals
* NH_4HS
* Ammonia
* Water

KEY FACTS

Diameter: 143,000 km
Average distance from Sun: 700 million km
Average distance from Earth: 629 million km
Day: 10 hours
Year: 12 Earth years
Moons: 95

JOVIAN WEATHER

Jupiter's magnetic fields produce aurorae at its poles — like on Earth, but much bigger.

The planet's internal heat source creates convection currents in the belts in its upper atmosphere.

Storms are common on Jupiter, including the Great Red Spot — which has been raging since at least 1830 and is now about the size of Earth. But it used to be twice as big, and scientists think the storm may soon pass and disappear.

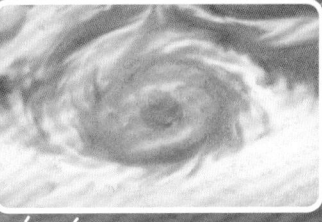

GREAT RED SPOT =
APPROX 1 EARTH

SCALE

EARTH   JUPITER

1,121.57%
OF EARTH

HUGE-PITER

The largest planet in the Solar System, Jupiter is 11x the diameter of Earth and has more than twice the mass of all the other seven planets combined!

GRAVITATIONAL SLINGSHOT

The gravity of a body can be used to alter the path and speed of a spacecraft — which can save on fuel. With its large mass, Jupiter has often been used in gravity assist manoeuvres.

ULYSSES
VOYAGER 1
PIONEER 10
CASSINI-HUYGENS

JUPITER
5.2 AU

EARTH — LAUNCH

- **1972** – Pioneer 10
- **1973** – Pioneer 11
- **1977** – Voyager 1
- **1977** – Voyager 2
- **1990** – Ulysses
- **1997** – Cassini-Huygens
- **2006** – New Horizons
- **2011** – Juno
- **2023** – JUICE

PIONEER 11
VOYAGER 2
NEW HORIZONS
JUNO
JUICE

EARTH
1 AU

SUN

JUPITER

- **1973** – Pioneer 10
- **1974** – Pioneer 11
- **1979** – Voyager 1
- **1979** – Voyager 2
- **1992** – Ulysses
- **2000** – Cassini-Huygens
- **2007** – New Horizons
- **2016** – Juno
- **2031** – JUICE (predicted)

ULYSSES

2004 – Made distant observations of Jupiter; 2009 last contact

Not to scale

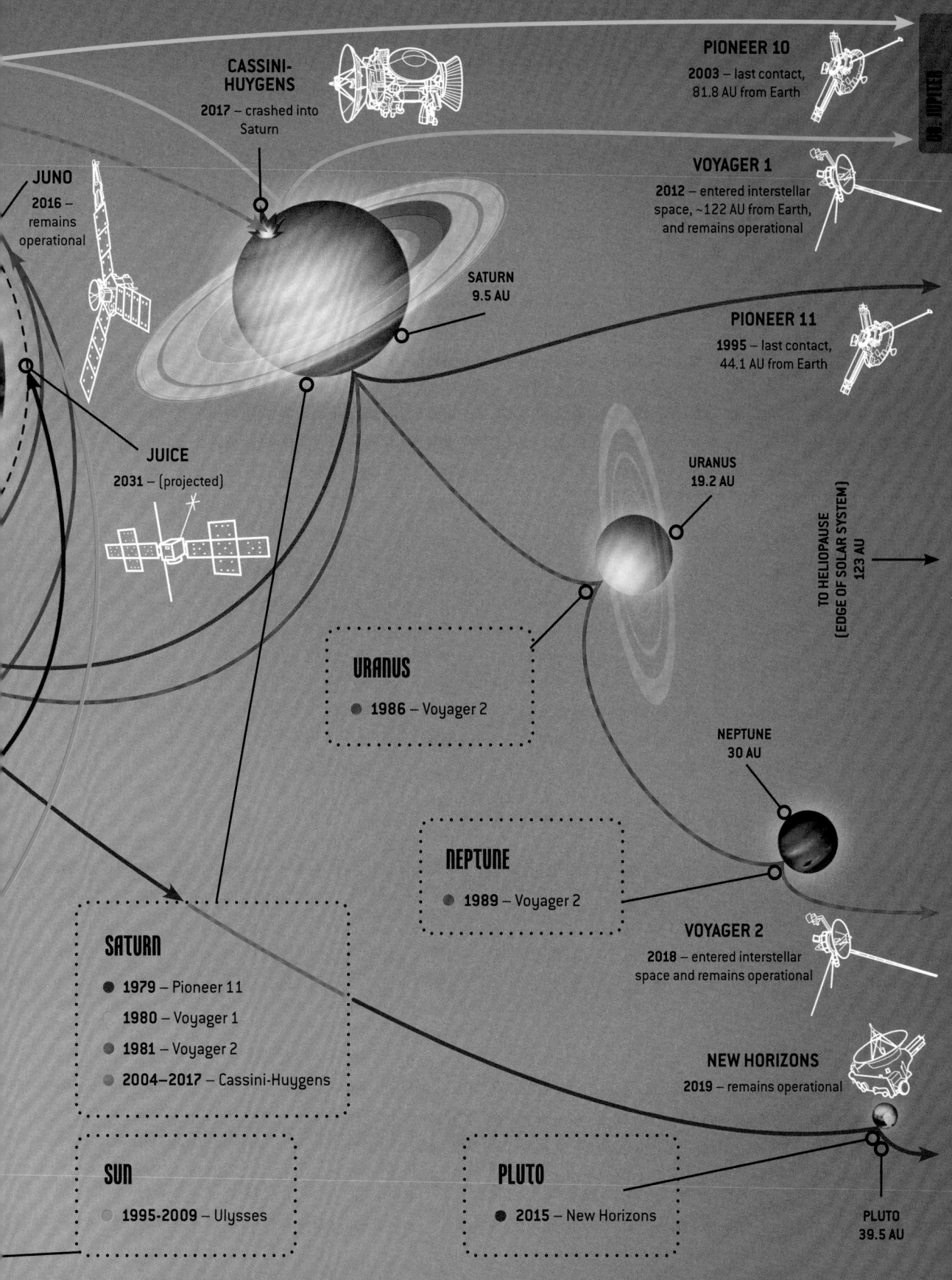

CASSINI-
HUYGENS

2017 – crashed into
Saturn

PIONEER 10

2003 – last contact,
81.8 AU from Earth

JUNO

2016 –
remains
operational

VOYAGER 1

2012 – entered interstellar
space, ~122 AU from Earth,
and remains operational

SATURN
9.5 AU

PIONEER 11

1995 – last contact,
44.1 AU from Earth

JUICE

2031 – (projected)

URANUS
19.2 AU

TO HELIOPAUSE
(EDGE OF SOLAR SYSTEM)
123 AU

URANUS

● 1986 – Voyager 2

NEPTUNE
30 AU

NEPTUNE

● 1989 – Voyager 2

VOYAGER 2

2018 – entered interstellar
space and remains operational

SATURN

● 1979 – Pioneer 11

1980 – Voyager 1

● 1981 – Voyager 2

● 2004–2017 – Cassini-Huygens

NEW HORIZONS

2019 – remains operational

SUN

● 1995-2009 – Ulysses

PLUTO

● 2015 – New Horizons

PLUTO
39.5 AU

THE MANY MOONS OF JUPITER

Giant Jupiter has 95 moons*

*New finds and updated data of location and distance are announced regularly (even as we wrote this book!)

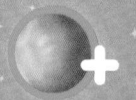

1. Metis
YEAR DISCOVERED: 1979
DISTANCE FROM JUPITER
(MILLION KM): 0.128

2. Adrastea
1979 | 0.129 MKM

3. Amalthea
1892 | 0.181 MKM

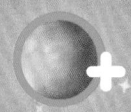

4. Thebe
1979 | 0.222 MKM

5. Io
1610 | 0.422 MKM

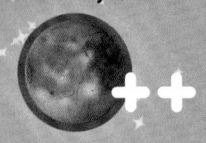

6. Europa
1610 | 0.671 MKM

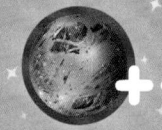

7. Ganymede
1610 | 1.070 MKM

8. Callisto
1610 | 1.883 MKM

9. Themisto
1975/2000 | 7.399 MKM

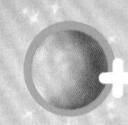

10. Leda
1974 | 11.146 MKM

11. Ersa
2018 | 11.401 MKM

12. S/2018 J 2
2018 | 11.420 MKM

13. Himalia
1904 | 11.441 MKM

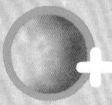

14. Pandia
2017 | 11.481 MKM

15. Lysithea
1938 | 11.701 MKM

16. Elara
1905 | 11.712 MKM

17. S/2011 J 3
2011 | 11.717 MKM

18. Dia
2000 | 12.260 MKM

19. S/2018 J 4
2018 | 16.329 MKM

20. Carpo
2003 | 17.042 MKM

21. Valetudo
2016 | 18.694 MKM

22. Euporie
2001 | 19.266 MKM

23. S/2003 J 18
2003 | 20.336 MKM

24. Eupheme
2003 | 20.769 MKM

25. S/2021 J 3
2021 | 20.777 MKM

26. S/2010 J 2
2010 | 20.793 MKM

27. S/2016 J 1
2016 | 20.803 MKM

28. Mneme
2003 | 20.821 MKM

29. Evanthe
2001 | 20.827 MKM

30. S/2003 J 16
2003 | 20.883 MKM

SCALE AND DISTANCE

JUPITER   METIS (Too small to see)   IO   EUROPA

KEY

CENTURY DISCOVERED:  ● 1600s ○ 1800s ● 1900s ○ 2000s

DIAMETER: ✛ More than 10 km ✛✛ More than 3,000 km

31. Harpalyke
2000 | 20.892 MKM

32. Orthosie
2001 | 20.901 MKM

33. Helike
2003 | 20.916 MKM

34. S/2021 J 2
2021 | 20.927 MKM

35. Praxidike
2000| 20.935 MKM

36. S/2017 J 3
2017 |20.941 MKM

37. S/2021 J 1
2021 | 20.955 MKM

38. S/2003 J 12
2003 | 20.963 MKM

39. S/2017 J 7
2017 | 20.965 MKM

40. Thelxinoe
2003 | 20.976 MKM

41. Thyone
2001 | 20.978 MKM

42. S/2003 J 2
2003 | 20.998 MKM

43. Ananke ✛
1951 | 21.035 MKM

44. S/2022 J 3
2022 | 21.048 MKM

45. Iocaste
2000 | 21.067 MKM

46. Hermippe
2001 | 21.109 MKM

47. S/2017 J 9
2017 | 21.769 MKM

48. Philophrosyne
2003 | 22.605 MKM

49. S/2016 J 3
2016 | 22.719 MKM

50. S/2022 J 1
2022 | 22.725 MKM

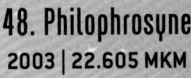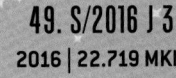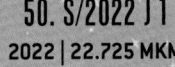

51. Pasithee
2001 | 22.847 MKM

52. S2017 J 8
2017 | 22.850 MKM

53. S/2021 J 6
2021 | 22.870 MKM

54. S/2003 J 24
2003 | 22.887 MKM

55. Eurydome
2001 | 22.899 MKM

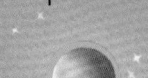

56. S/2011 J 2
2011 | 22.909 MKM

57. S/2003 J 4
2003 | 22.927 MKM

58. Chaldene
2000 | 22.931 MKM

59. S/2017 J 2
2017 | 22.953 MKM

60. Isonoe
2000 | 22.981 MKM

SOURCE: NASA Solar System Dynamics, https://ssd.jpl.nasa.gov/sats/elem/

Jupiter and Jupiter and moons shown to scale in size but at 1/6 of actual distance.

GANNYMEDE

CALLISTO

THE MANY MOONS - CONTINUED

61. S/2022 J 2
2022 | 23.014 MKM

62. S/2021 J 4
2021 | 23.020 MKM

63. Kallichore
2003 | 23.022 MKM

64. Erinome
2000 | 23.033 MKM

65. Kale
2001 | 23.053 MKM

66. Eirene
2003 | 23.056 MKM

67. Aitne
2001 | 23.064 MKM

68. Eukelade
2003 | 23.067 MKM

69. Arche
2002 | 23.098 MKM

70. Taygete
2000 | 23.108 MKM

71. S/2016 J 4
2016 | 23.114 MKM

72. S/2011 J 1
2011 | 23.125 MKM

73. Carme
1938 | 23.144 MKM

74. Herse
2003 | 23.151 MKM

75. S/2003 J 19
2003 | 23.156 MKM

76. S/2010 J 1
2010 | 23.190 MKM

77. S/2003 J 9
2003 | 23.199 MKM

78. S/2017 J 5
2017 | 23.206 MKM

79. S/2017 J 6
2017 | 23.245 MKM

80. Kalyke
200 | 23.303 MKM

81. Hegemone
2003 | 23.349 MKM

82. S/2018 J 3
2018 | 23.400 MKM

83. S/2021 J 5
2021 | 23.415 MKM

84. Pasiphae
1908 | 23.468 MKM

85. Sponde
2001 | 23.543 MKM

86. S/2003 J 10
2003 | 23.576 MKM

87. Megaclite
2000 | 23.645 MKM

88. Cyllene
2003 | 23.655 MKM

89. Sinope
1914 | 23.684 MKM

90. S/2017 J 1
2017 | 23.745 MKM

91. Aoede
2003 | 23.778 MKM

92. Autonoe
2001 | 23.793 MKM

93. Callirrhoe
1999 | 23.796 MKM

94. S/2003 J 23
2003 | 23.829 MKM

95. Kore
2003 | 24.205 MKM

SCALE AND DISTANCE - CONTINUED

TO JUPITER
←

LOST MOONS

How do you lose a moon?

THEMISTO

One moon of Jupiter was first observed in 1975 — and then lost until 2000! In 2002, with its path confirmed, it was named Themisto.

CRASH

Moons might collide and be destroyed or create other, smaller moons!

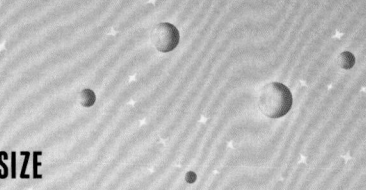

SIZE

Many of these moons are very small and faint, so difficult to spot.

DISTANCE

They're also a long way from the planet, meaning they take a long time to complete an orbit.

COST

So big telescopes must observe for a long time to confirm an orbital path. That is not always practical with limited resources.

ECCENTRICITY

Orbits may be eccentric, steeply inclined and/or affected by the gravity of other moons.

LACK OF DATA

So far, we have little observational data for three recently discovered moons of Jupiter — S/2022 J 1, S/2022 J 2 and S/2022 J 3 — so they're at risk of being lost!

It's 7,393,216 km from Jupiter to Themisto, its ninth moon.

TO THEMISTO

GALILEAN MOONS

The four largest moons of Jupiter, discovered by Galileo in 1609–10, are very different from one another.

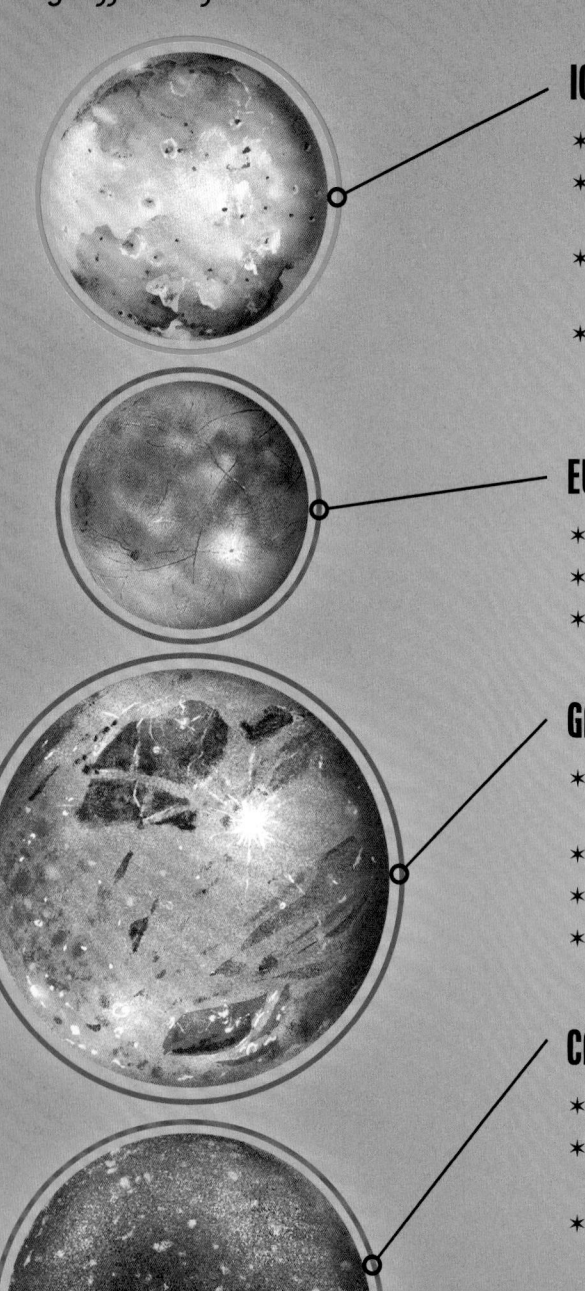

IO

* Most geologically active body in the Solar System
* Surface temperatures raised and lowered by 100°C due to tidal heating — the effect of Jupiter's gravity!
* More than 100 volcanoes, with gas and dust erupting 250 km above surface!
* Coloured surface due to compounds and forms of sulphur in eruptions

EUROPA

* Crust of water ice covers liquid salt water ocean 1,000 km deep
* Many consider this the most likely place to find alien life!
* See **LIFE ON EUROPA?** 122

GANYMEDE

* Largest moon in the Solar System — and bigger than Mercury and Pluto
* Flat surface, no features above 1 km high
* Composed half of silicate rock and half of water ice
* Has its own magnetic field

CALLISTO

* Surface heavily cratered, with craters in wide range of sizes
* This suggests that, unlike other Galilean moons, the surface of Ganymede has changed little in about 4 billion years
* Features include eight multi-ringed basins, such as 3,000 km-wide Valhalla

SCALE AND DISTANCE – CONTINUED

TO JUPITER ←

ORBITAL RESONANCE

The three inner Galilean moons are in 4:2:1 orbital resonance with each other: their gravitational effect on each other means that every 7.155 Earth days…

JUPITER

…Io completes 4 orbits,

Europa completes 2 orbits,

and Ganymede completes 1 orbit.

In the same period, Callisto completes 42.87% of an orbit.

TEN LARGEST MOONS IN THE SOLAR SYSTEM

With Mercury and a selection of dwarf planets shown for scale.

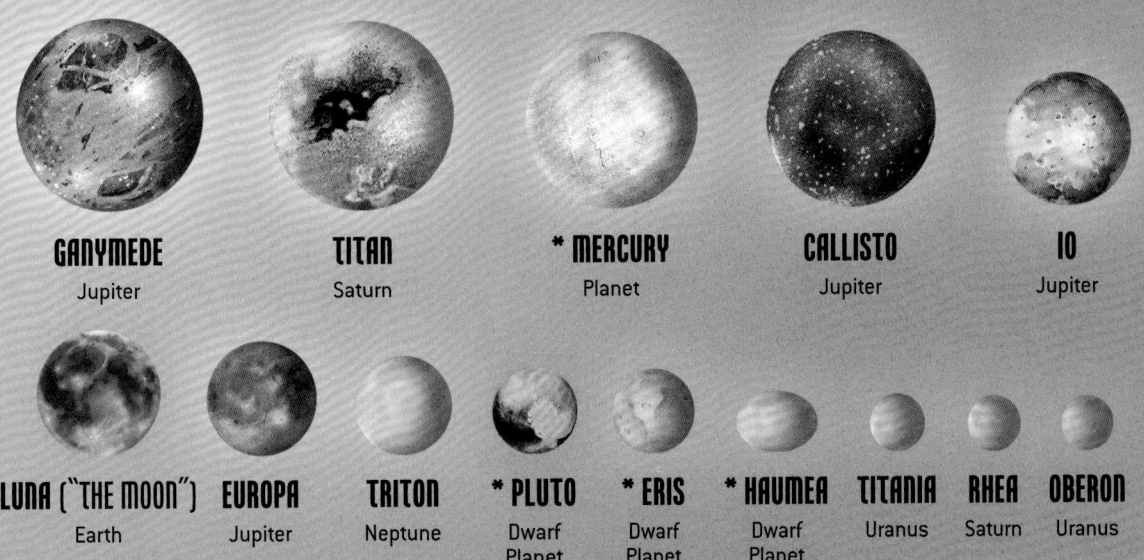

GANYMEDE
Jupiter

TITAN
Saturn

* MERCURY
Planet

CALLISTO
Jupiter

IO
Jupiter

LUNA (“THE MOON”)
Earth

EUROPA
Jupiter

TRITON
Neptune

* PLUTO
Dwarf Planet

* ERIS
Dwarf Planet

* HAUMEA
Dwarf Planet

TITANIA
Uranus

RHEA
Saturn

OBERON
Uranus

We're now 5,727,678 km from Jupiter, only another 1,665,538 km to Themisto.

TO THEMISTO →

CLEAN-UP

Jupiter's powerful gravitational influence attracts objects that might otherwise be hazardous to Earth.

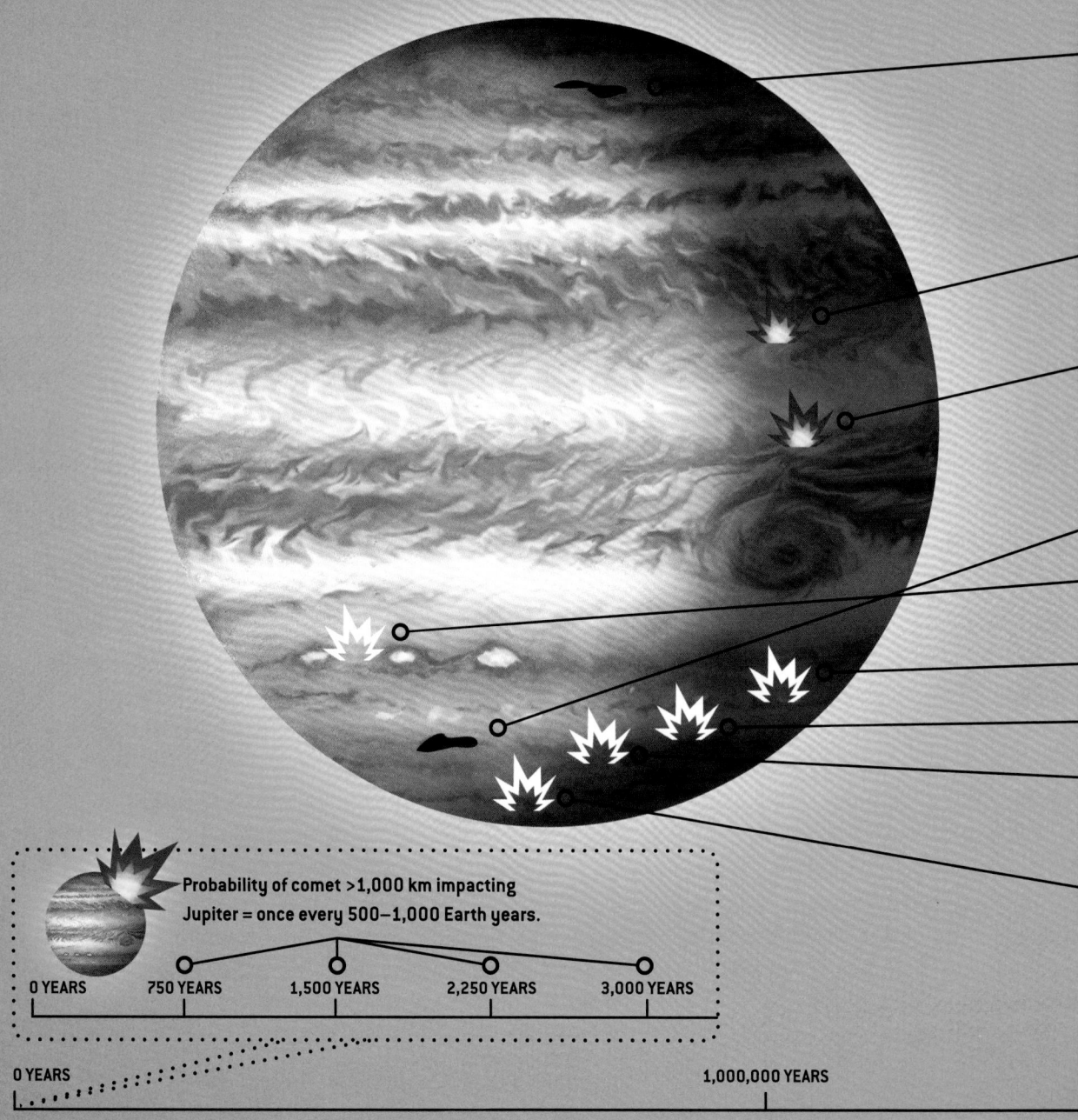

Probability of comet >1,000 km impacting Jupiter = once every 500–1,000 Earth years.

0 YEARS	750 YEARS	1,500 YEARS	2,250 YEARS	3,000 YEARS

0 YEARS	1,000,000 YEARS

SCALE AND DISTANCE – CONTINUED

TO JUPITER
←

OBSERVED IMPACTS

16–22 JULY 1994

Comet Shoemaker-Levy 9 (1,800 m in diameter) broke apart and hit Jupiter in a series of impacts.

This first suggested the planet's role in "cleaning up" the debris of the Solar System.

7 DECEMBER 1995

Galileo probe

GALILEO PROBE
(Before separation)

21 SEPTEMBER 2003

Galileo spacecraft

19 JULY 2009

Unnamed object (estimated to be 200–500 m), resulting in temporary black "scar" in Jupiter's atmosphere 8,000 km long.

3 JUNE 2010

Unnamed object (8–13 m), resulting in a two-second flash observed by astronomers on Earth.

It has since been estimated that Jupiter is impacted by objects of similar size several times each Earth year.

20 AUGUST 2010

Unnamed object (10 m), with observed brief flash.

10 SEPTEMBER 2012

Unnamed object (30 m), with observed brief flash.

17 MARCH 2016

Unnamed object (15 m), with observed brief flash.

26 MAY 2017

Unnamed object (13 m), with observed brief flash.

Probability of comet > 1,000 km impacting Earth = once every 2–4 million Earth years.

2,000,000 YEARS

3,000,000 YEARS

Themisto is only 8 km in diameter while the tenth moon, Leda, is twice the size at 16 km!

THEMISTO (Too small to see)

TO LEDA →

ATMOSPHERES

What you'd be breathing on the surface of each of the eight planets.

ROCKY PLANETS

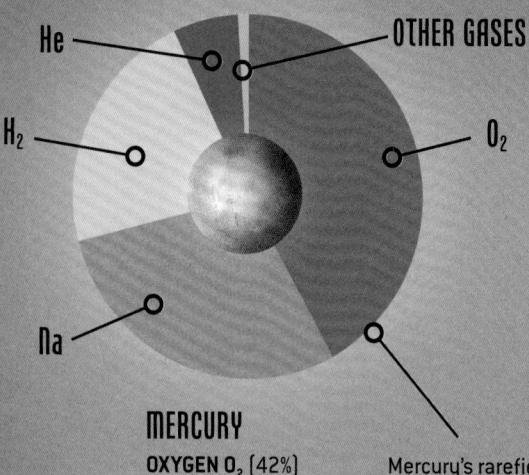

He

H_{2}

OTHER GASES

O_{2}

Na

MERCURY
OXYGEN O_{2} (42%)
SODIUM Na (29%)
HYDROGEN H_{2} (22%)
HELIUM He (6%)
OTHER GASES (1%)

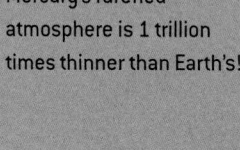

Mercury's rarefied atmosphere is 1 trillion times thinner than Earth's!

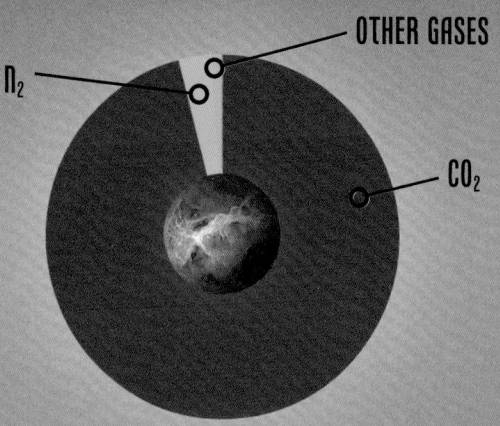

N_{2}

OTHER GASES

CO_{2}

VENUS
CARBON DIOXIDE CO_{2} (96%)
NITROGEN N_{2} (3%)
OTHER GASES (1%)

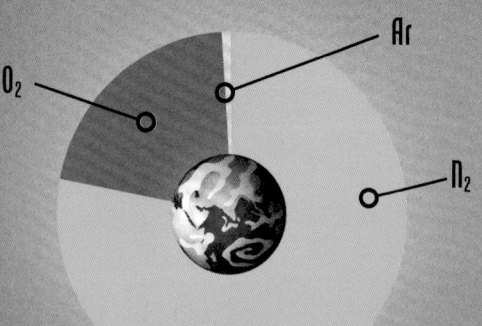

O_{2}

Ar

N_{2}

EARTH
NITROGEN N_{2} (78%)
OXYGEN O_{2} (21%)
ARGON Ar ($\cong$ 1%)
OTHER GASES (<1%)

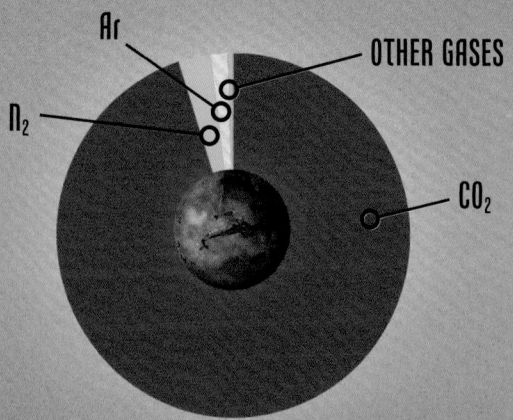

Ar

OTHER GASES

N_{2}

CO_{2}

MARS
CARBON DIOXIDE CO_{2} (95%)
NITROGEN N_{2} (3%)
ARGON Ar (1.5%)
OTHER GASES (0.5%)

SCALE AND DISTANCE – CONTINUED

TO JUPITER
←

GAS GIANTS

OTHER GASES

He

H₂

JUPITER
HYDROGEN H₂ (90%)
HELIUM He (≅10%)
OTHER GASES (<1%)

SATURN
HYDROGEN H₂ (96%)
HELIUM He (3%)
OTHER GASES (1%)

OTHER TRACE HYDROCARBONS

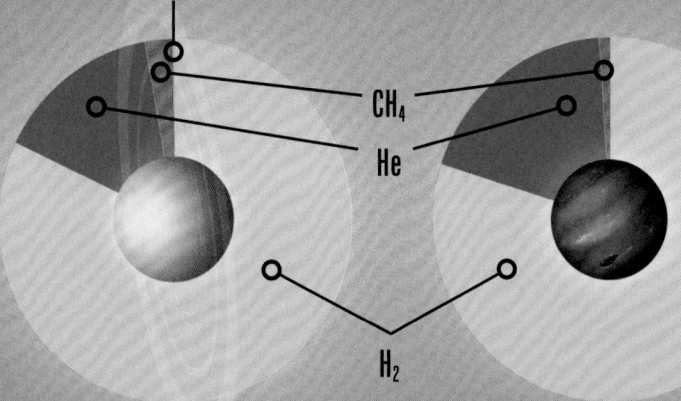

CH₄

He

H₂

URANUS
HYDROGEN H₂ (83%)
HELIUM He (15%)
METHANE CH₄ (2.5%)
OTHER TRACE HYDROCARBONS (<1%)

NEPTUNE
HYDROGEN H₂ (80%)
HELIUM He (19%)
METHANE CH₄ (≅1%)

NB: Atmospheric pressures vary greatly. Mercury's atmosphere is so thin it is called an exosphere.

Planets not to scale

We're now 9,546,130 km away from Jupiter, another 1,641,651 km unti Leda.

TO LEDA

121

LIFE ON EUROPA?

Jupiter's moon may be the most likely place in the Solar System we'll find life as we know it.

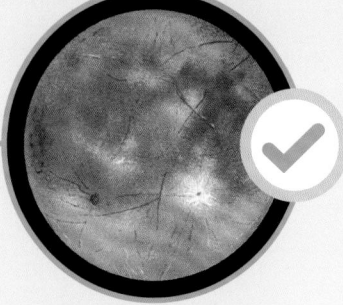

EUROPA (MOON OF JUPITER)

✓ Salty ocean under its frozen crust (15–20 km thick) contains more water than Earth's oceans.

✓ Minerals detected on surface are — on Earth — associated with organic matter. Europa's ocean may have a rocky floor, from which chemicals can be released into the water.

✓ Smooth ice surface suggests it is new — heat melting and refreezing the ice. Europa seems to be squeezed by the gravity of nearby Jupiter, in a process called "tidal flexing" that produces heat, especially at boundary between liquid ocean and ice crust. Europa's surface is also blasted by radiation from Jupiter which might fuel life in the ocean below.

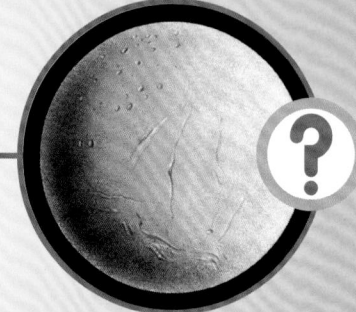

ENCELADUS (MOON OF SATURN)

✓ Vast oceans of water under the ice crust (30–40 km thick).

✓ Carbon, nitrogen and hydrogen in the oceans — all vital for life, and plumes ejected into space include organic molecules. One micro-organism, *Methanothermococcus okinawensis,* can thrive in the same conditions on Earth — so life is possible.

? Enceladus ejects plumes of salty water suggesting hydrothermal activity within the ocean beneath its icy surface. Such hydrothermal energy is enough to support life deep in the oceans on Earth.

MARS

✓ Mars is just within the habitable or "goldilocks" zone, with large amounts of water underground and some ice (and possibly liquid water) on surface. See **VENUS > KEEP YOUR DISTANCE** 📖 36

✓ Martian soil contains chemicals that could support life.

✗ Mars is very cold and its thin, oxidising atmosphere would not protect life on the planet's surface from intense ultraviolet radiation. Studies are focused on locating evidence of past Martian life rather than there being life on Mars now.

SCALE AND DISTANCE – CONTINUED

TO JUPITER ←

KEY

💧 WATER   ⚛ CHEMISTRY   🔥 ENERGY

VENUS

💧 ❓ Although the surface is too hot for liquid water, milder temperatures are found in the cloud layers surrounding the planet — which include water vapour.

⚛ ❓ The atmosphere is rich in chemicals, apparently produced by volcanic activity on the planet's surface. These include carbonyl sulphide which is usually produced organically — perhaps suggesting life.

🔥 ❓ Micro-organisms in cloud layer might absorb ultraviolet light from the Sun — and research suggests something in the atmosphere is absorbing UV light and affecting the brightness of Venus.

TITAN (MOON OF SATURN)

💧 ❓ There may be liquid water underground, but the surface of Titan has liquid methane and ethane — which might be capable of supporting life.

⚛ ❓ Atmosphere rich in organic compounds. Experiments have shown that adding ultraviolet to similar chemicals can produce complex molecules — suggesting basic stages of life would be possible if there was an energy source.

🔥 ✖ Titan is extremely cold at -179 °C. It's unlikely there is life there now, but studying it may help us understand conditions on Earth before life began.

CERES (DWARF PLANET)

💧 ❓ Thin atmosphere of water vapour, and there may be some frost on the surface.

⚛ ❓ Organic compounds detected on the surface.

🔥 ✖ Temperatures range from -163.5 to -118°C.

Leda is an average of 7,393,216 km from Jupiter.

LEDA (Too small to see)   **HIMALIA** (Too small to see)   **TO ERSA**

COMETS

Lumps of rock, dust and ice in highly elliptical orbits that sublime close to the Sun.

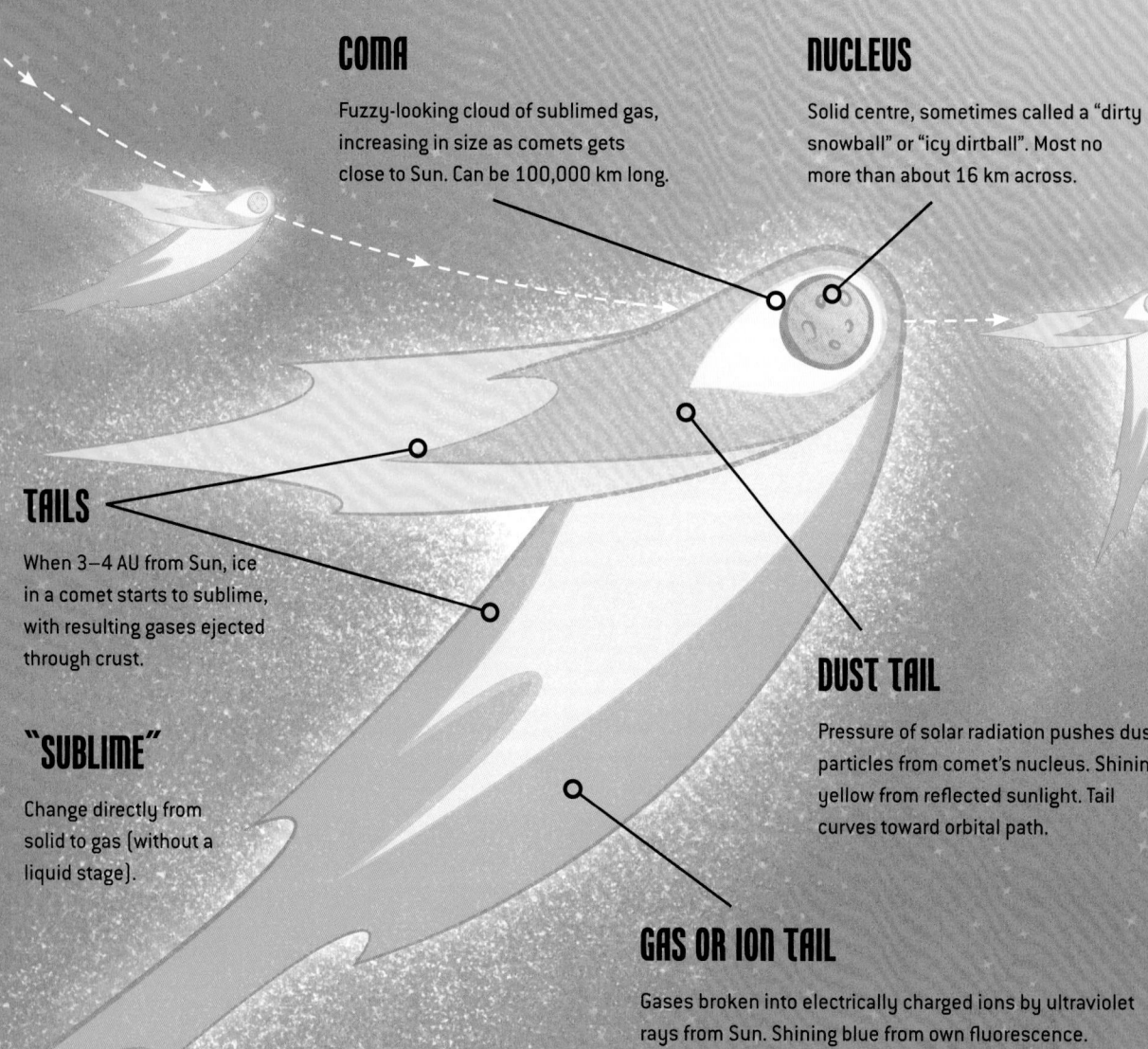

COMA

Fuzzy-looking cloud of sublimed gas, increasing in size as comets gets close to Sun. Can be 100,000 km long.

NUCLEUS

Solid centre, sometimes called a "dirty snowball" or "icy dirtball". Most no more than about 16 km across.

TAILS

When 3–4 AU from Sun, ice in a comet starts to sublime, with resulting gases ejected through crust.

"SUBLIME"

Change directly from solid to gas (without a liquid stage).

DUST TAIL

Pressure of solar radiation pushes dust particles from comet's nucleus. Shining yellow from reflected sunlight. Tail curves toward orbital path.

GAS OR ION TAIL

Gases broken into electrically charged ions by ultraviolet rays from Sun. Shining blue from own fluorescence. Points directly away from direction of Sun as is swept back by particles in solar wind.

SCALE AND DISTANCE – CONTINUED

TO JUPITER	ERSA (Too small to see)	PANDIA (Too small to see)	LYSITHEA (Too small to see)	ELARA (Too small to see)

SHORT-PERIOD COMETS

Orbital periods of <200 Earth years.

JUPITER FAMILY

JUPITER FAMILY comets have orbital periods <20 Earth years and aphelia at ≅ 5 AU.

As comets often have highly elliptical or even open-ended orbits, they can appear anywhere in the sky above Earth, not only in the plain of the ecliptic (as with planets).

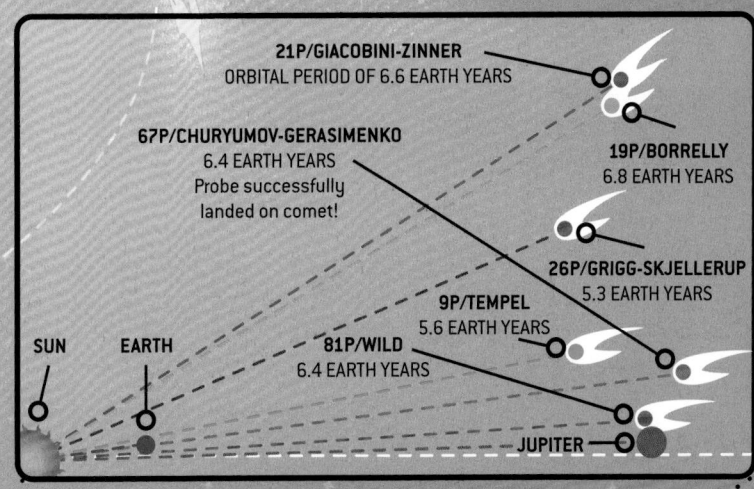

21P/GIACOBINI-ZINNER
ORBITAL PERIOD OF 6.6 EARTH YEARS

67P/CHURYUMOV-GERASIMENKO
6.4 EARTH YEARS
Probe successfully landed on comet!

19P/BORRELLY
6.8 EARTH YEARS

26P/GRIGG-SKJELLERUP
5.3 EARTH YEARS

9P/TEMPEL
5.6 EARTH YEARS

81P/WILD
6.4 EARTH YEARS

SUN   EARTH

JUPITER

1P/HALLEY
75.32 EARTH YEARS

KUIPER BELT INNER EDGE

SUN   EARTH   JUPITER   SATURN

ECLIPTIC

HALLEY FAMILY

HALLEY FAMILY comets have periods 20–200 Earth years (thought to be Oort Cloud objects captured by gravity of outer planets).

We've sent six probes to a single comet of this type: **1P/Halley**.

LONG-PERIOD COMETS

Orbital periods of >200 Earth years

Some can have periods of 1,000s of Earth years! Aphelia in Oort Cloud. For example: **Hyakutake**.

HYAKUTAKE
≅ 70,000 EARTH YEARS

ECLIPTIC

SUN

KEY

- – – INNER EDGE OF OORT CLOUD
- – – – KUIPER BELT SCATTERED DISK
- ⬡ HELIOPAUSE

Carpo is another 3,780,291 km away from Jupiter (another four pages)!

DIA
(Too small to see)

TO CARPO →

ZODIACAL LIGHT

The dust left behind by comets and asteroids can have a striking effect in Earth's sky.

Western sky after sunset during spring in the northern hemisphere

ZODIACAL LIGHT is a faint, triangular glow in Earth's night sky.

It's the result of sunlight being scattered by a thin, pancake-shaped cloud of tiny particles of dust in the same plain as the ecliptic and the constellations of the zodiac.

CONSTELLATION — PISCES

LINE OF ECLIPTIC

SUNLIGHT REFLECTED ON DUST PARTICLES 10 TO 300 MILLIONTHS OF A METRE IN SIZE!

VENUS

TO SUN

ECLIPTIC

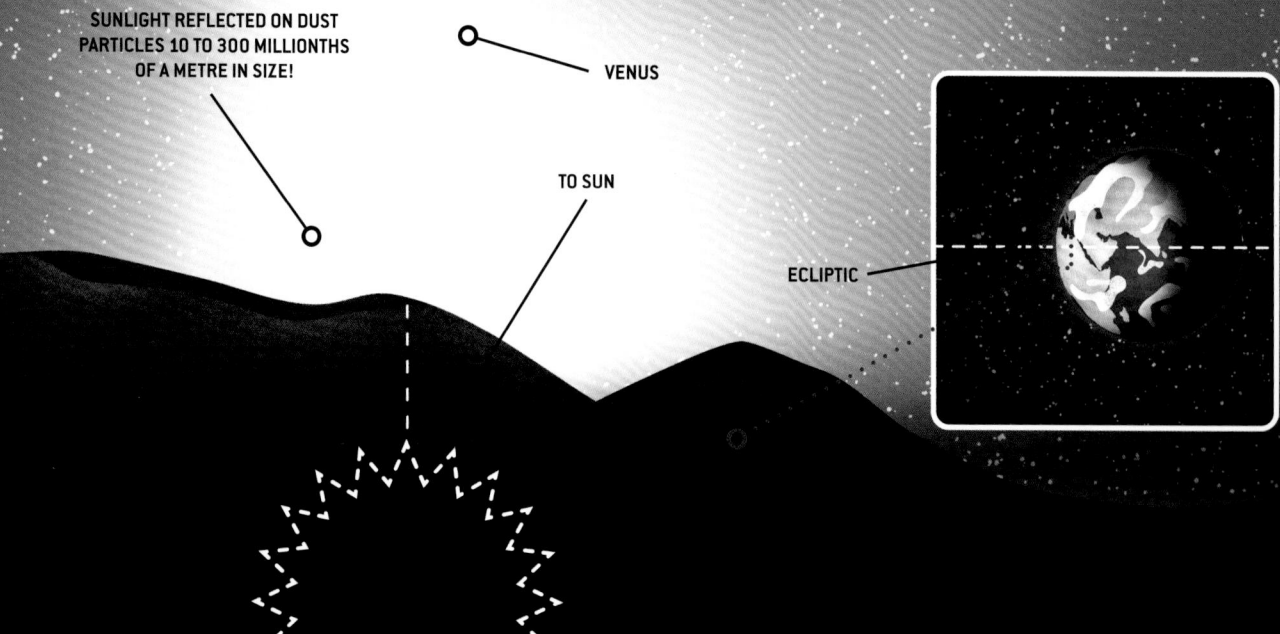

SCALE AND DISTANCE — CONTINUED

TO JUPITER

Zodiacal light is best seen when the ecliptic is at a steep angle to Earth's horizon — after sunset in spring and before sunrise in autumn. But it is easily outshone by moonlight or light pollution on Earth.

The dust is left behind by passing comets and the collisions of asteroids.

The tiniest particles of dust are blown away by solar wind. The cloud is maintained by more passing comets and asteroid collisions.

GEGENSCHEIN

Bright oval caused by back-scattered sunlight on dust cloud. It appears at the "antisolar point" in the night sky, directly opposite the (unseen) Sun's position from the observer's perspective.

It is brighter than zodiacal light because it is seen at full phase to the Sun — just as the Moon is brightest at full Moon.

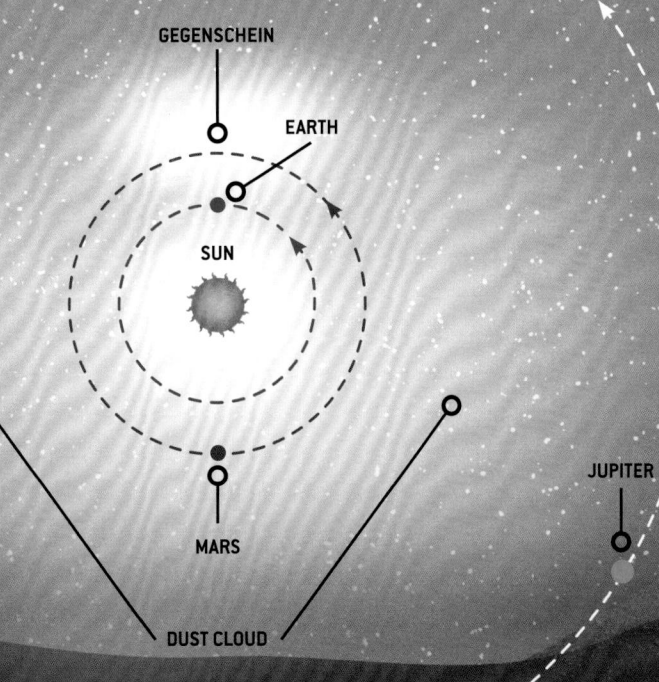

GEGENSCHEIN

EARTH

SUN

JUPITER

MARS

DUST CLOUD

Still another 1,871,065 km (two pages) until Carpo.

TO CARPO

TROJANS

More than a million asteroids share Jupiter's orbital path around the Sun.

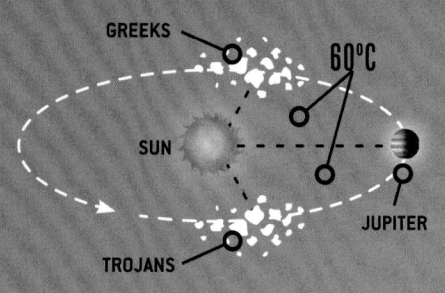

GREEKS

60°C

SUN

JUPITER

TROJANS

JUPITER TROJANS (& GREEKS)

More than 7,000 individual Trojan asteroids have been identified but the Greek and Trojan groups are each thought to contain:

* 600,000 asteroids larger than 1 km
* 200,000 asteroids larger than 2 km

That's about the same as in the main Asteroid Belt!

Trojans sit at **LAGRANGE POINTS** in the orbital path. See **MARS > SPACE JARGON – II** [92]

TRAILING "TROJANS"

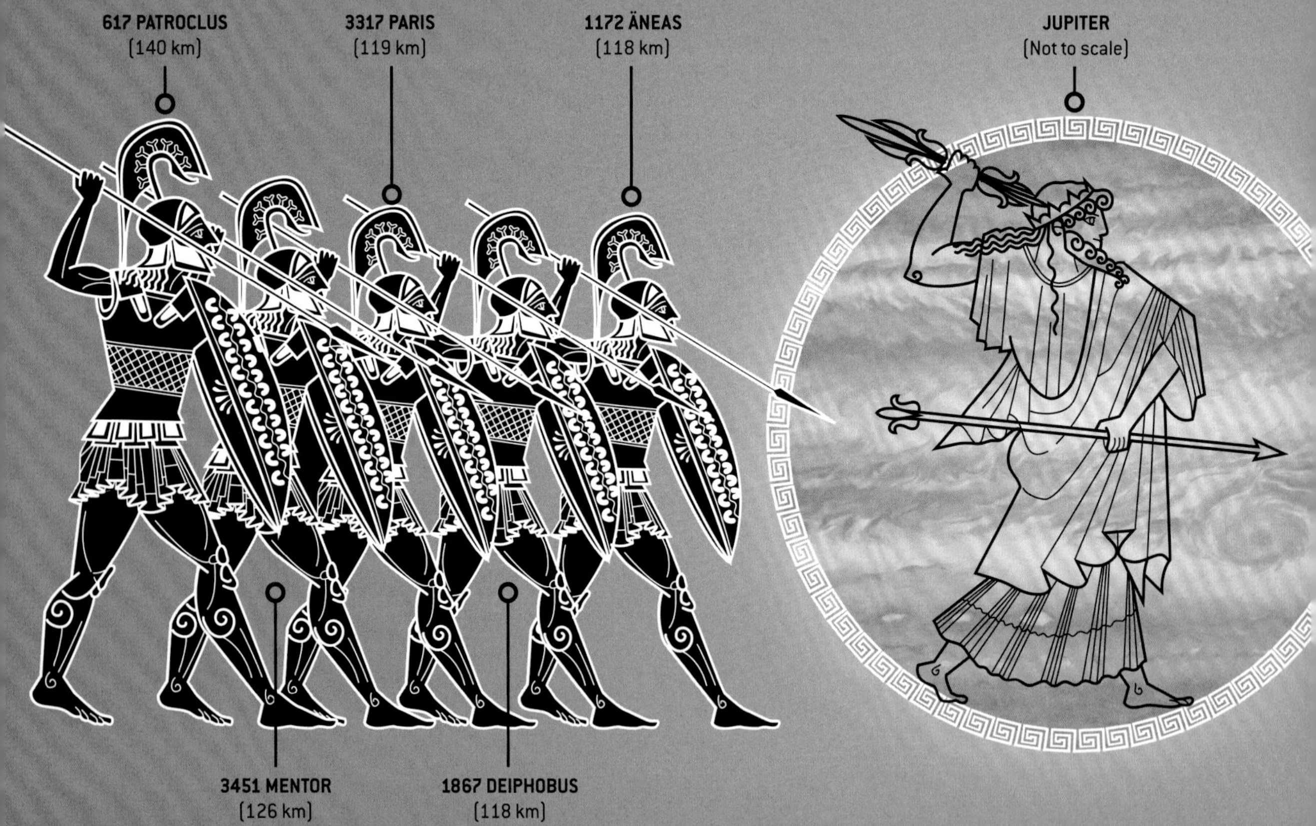

617 PATROCLUS
(140 km)

3317 PARIS
(119 km)

1172 ÄNEAS
(118 km)

JUPITER
(Not to scale)

3451 MENTOR
(126 km)

1867 DEIPHOBUS
(118 km)

SCALE AND DISTANCE – CONTINUED

TO JUPITER
←

EVENLY MATCHED

The 10 largest Trojans and Greeks.

The "Greek" and "Trojan" asteroids are mostly named after characters from each side in the mythical Trojan War, but Hektor was a Trojan in that war and Patroclus a Greek. The naming convention was established after they'd been discovered!

624 HEKTOR
(225 km) – has a 12 km satellite of its own, Skamandrios

LEADING "GREEKS"

1143 ODYSSEUS
(115 km)

588 ACHILLES
(130 km) – the first Trojan discovered, in 1906

1437 DIOMEDES
(118 km)

911 AGAMEMNON
(131 km)

OTHER TROJANS

"Trojan Asteroids" are specifically those in the same orbital path as Jupiter - including both Greek and Trojan groups. But there are other trojans in the Solar System...

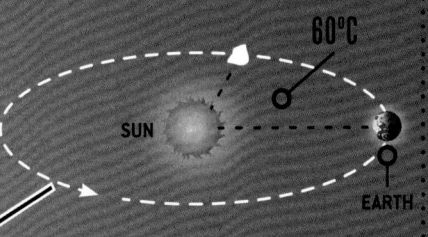

60°C

SUN

EARTH

EARTH TROJAN

* 1 known leading asteroid, 2010 TK7 (300 m)

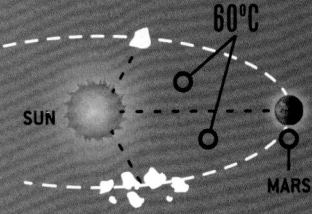

60°C

SUN

MARS

MARS TROJANS

* 1 known leading asteroid
* 6 known trailing asteroids, plus one potential candidate

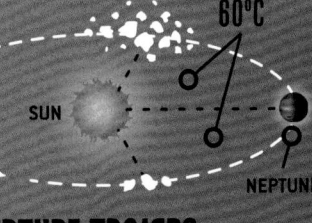

60°C

SUN

NEPTUNE

NEPTUNE TROJANS

* 19 known leading asteroids
* 3 known trailing asteroids

There are another 62 moons, and if shown would continue for 12 more pages.

CARPO
(Too small to see)

TO MORE MOONS

09. SATURN

UPPER ATMOSPHERE

✴ Temperature of upper atmosphere −183°C

✴ Similar to Jupiter except for lack of ammonia and less helium due to lower temperatures

✴ Similar to Jupiter with zones, belts and spots, but hidden by a high-altitude, cream-coloured haze

EQUATOR

✴ Eastward equatorial winds of 500 m/s!

CORE

✴ Thought to be similar to Jupiter

LIQUID HYDROGEN

✴ 1,000 km down, pressure high enough to turn gas to liquid hydrogen

METALLIC HYDROGEN & HELIUM

✴ 90,000 km down, pressure high enough for liquid metallic hydrogen and helium, which generates magnetic field

RINGS

Rings of ice and rock visible by telescope from Earth:

✴ 7,000 km to 80,000 km from Saturn's equator

✴ From 10 m to 1 km thick

See RINGS 134

KEY FACTS

Diameter: 120,549 km
Average distance from Sun: 1,430 million km
Average distance from Earth: 1,197 million km
Day: ~10.6 hours
Year: 29.5 Earth years
Moons: 146

LIGHT PLANET

Saturn's bulk density is so low it could float in water.

HOT PLANET

Saturn radiates 2.5x more energy into space than it receives from the Sun.

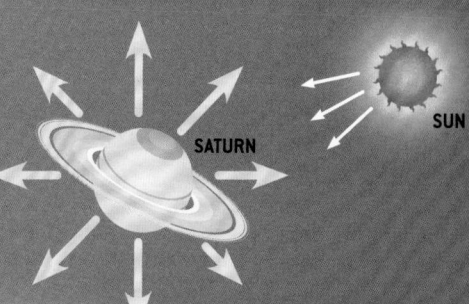

SATURN

SUN

SCALE

EARTH

SATURN

945.48%
OF EARTH

VISIONS OF SATURN

How technology has changed our view of this extraordinary planet.

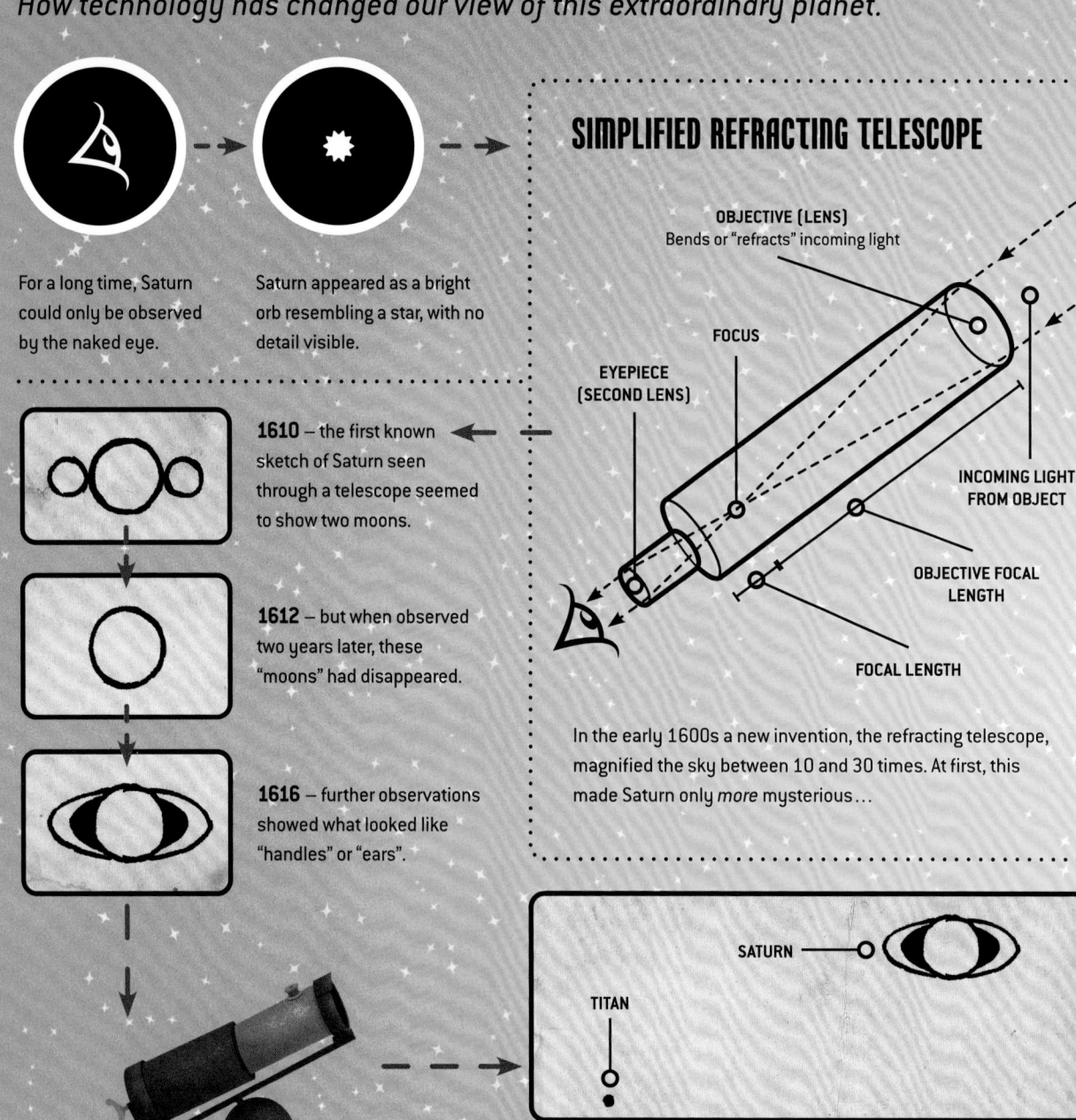

For a long time, Saturn could only be observed by the naked eye.

Saturn appeared as a bright orb resembling a star, with no detail visible.

SIMPLIFIED REFRACTING TELESCOPE

OBJECTIVE (LENS)
Bends or "refracts" incoming light

FOCUS

EYEPIECE (SECOND LENS)

INCOMING LIGHT FROM OBJECT

OBJECTIVE FOCAL LENGTH

FOCAL LENGTH

1610 – the first known sketch of Saturn seen through a telescope seemed to show two moons.

1612 – but when observed two years later, these "moons" had disappeared.

1616 – further observations showed what looked like "handles" or "ears".

In the early 1600s a new invention, the refracting telescope, magnified the sky between 10 and 30 times. At first, this made Saturn only *more* mysterious…

Longer telescopes with wider, better-made lenses increased magnification.

SATURN

TITAN

1655 – 50x magnification: discovery of Titan, the first known of Saturn's moons

1656 – 100x magnification: realisation that changing "ears" are rings, their angle inclined to Earth.

REFLECTING TELESCOPE

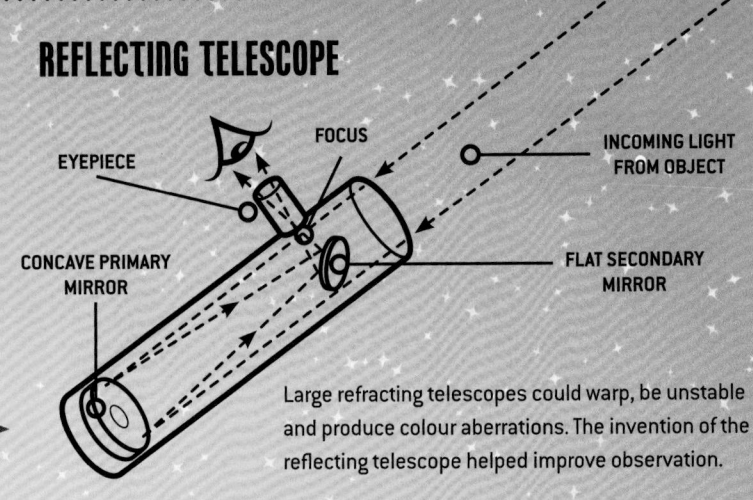

EYEPIECE

FOCUS

INCOMING LIGHT FROM OBJECT

CONCAVE PRIMARY MIRROR

FLAT SECONDARY MIRROR

Large refracting telescopes could warp, be unstable and produce colour aberrations. The invention of the reflecting telescope helped improve observation.

With improved telescopes (both refracting and reflecting), we could spot gaps or "divisions" in the rings, and coloured bands across the surface of the planet. But images could still be indistinct and lacked detail.

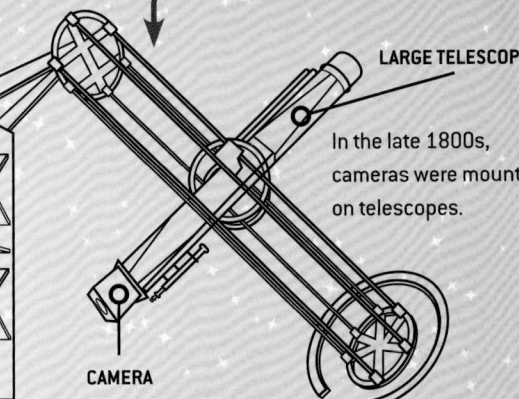

LARGE TELESCOPE

In the late 1800s, cameras were mounted on telescopes.

CAMERA

1899 – Long exposure photographs captured more light from objects otherwise too faint to see.

Comparing photographs taken at different times showed these faint objects to be more of Saturn's moons.

1966 – While Saturn's rings were angled edge-on to Earth (and almost invisible), observers discovered a tenth moon of Saturn. That same year, they spotted another object that proved to be an eleventh moon.

1980 – Voyager 1 is the first of many space probes to reach Saturn and immediately discovers three more moons.

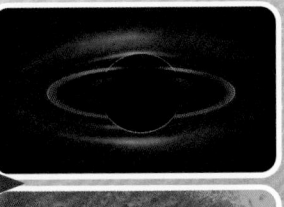

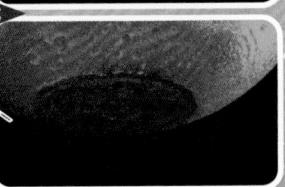

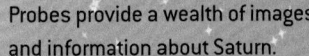

Probes provide a wealth of images and information about Saturn.

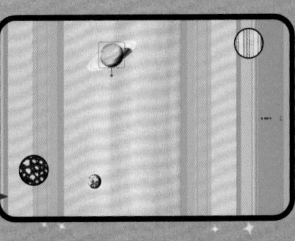

Analysing the data collected can take years. For example, observations of Saturn made in 2004 were discovered, in 2019, to include 20 previously unknown moons.

RINGS – I

The striking bands of ice and rock that orbit Saturn.

Saturn and its rings have been stretched into lines, showing the average distance from surface of Saturn.

BRIGHT RINGS

Bright, most easily visible part of ring system extends 7,000 to 80,000 km from Saturn's equator, and ranges from 10 m to 1 km thick.

COLOMBO GAP/
TITAN RINGLET

CASSINI
DIVISION

A
RING

B
RING

C
RING

D
RING

SATURN

MAXWELL
GAP/RINGLET

ICE AND ROCK

Composed of ice and rock, from small grains to boulders a few metres across. Each particle moves in its own orbit round Saturn.

EARTH
FOR SCALE

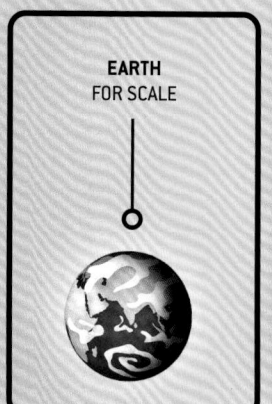

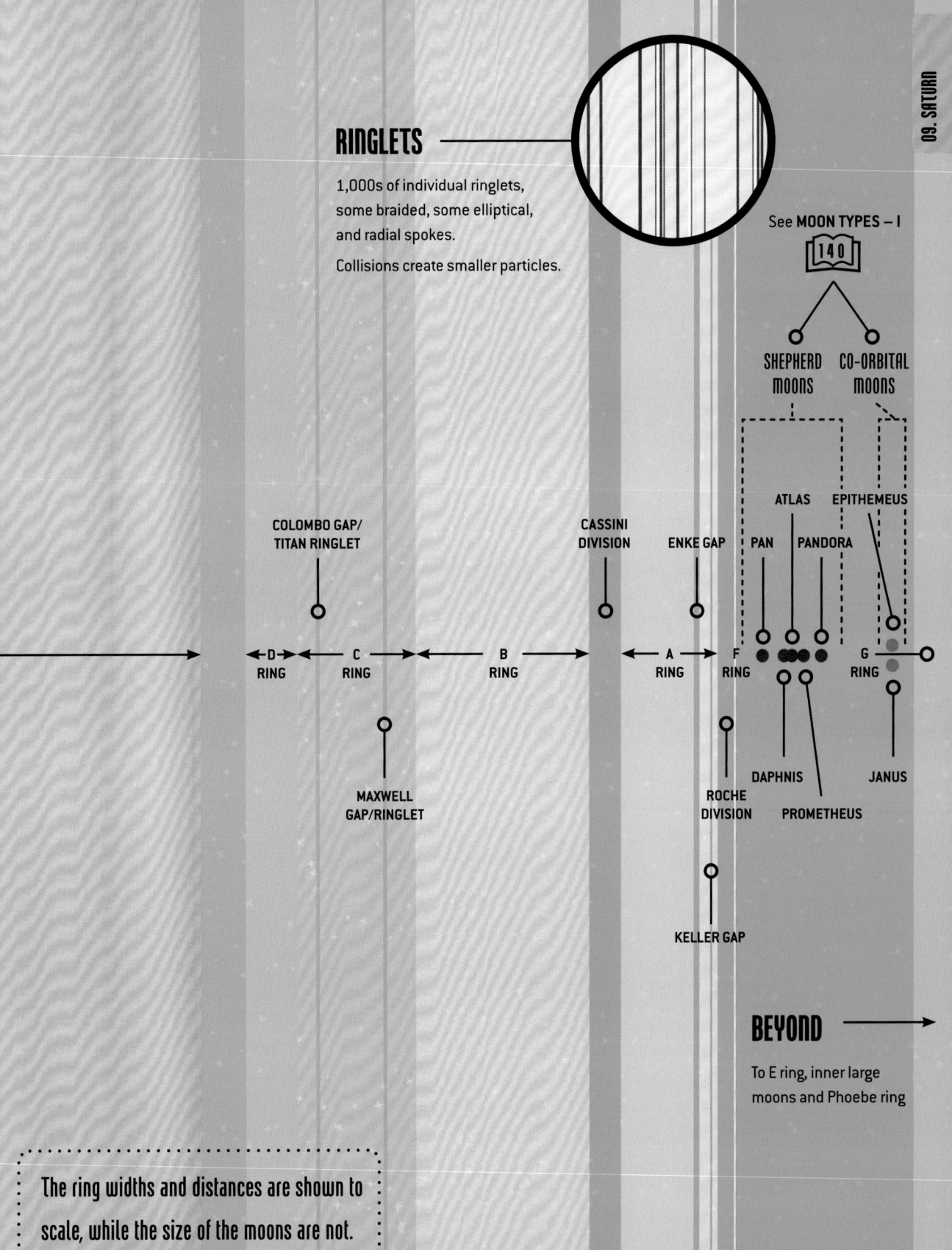

RINGLETS

1,000s of individual ringlets, some braided, some elliptical, and radial spokes.

Collisions create smaller particles.

See **MOON TYPES – I**

[140]

SHEPHERD MOONS

CO-ORBITAL MOONS

ATLAS

EPITHEMEUS

COLOMBO GAP/ TITAN RINGLET

CASSINI DIVISION

ENKE GAP

PAN

PANDORA

◄ D ► ◄━━ C ━━► ◄━━━━━ B ━━━━━► ◄━ A ━► F ◄━ RING RING RING RING RING

G RING

MAXWELL GAP/RINGLET

ROCHE DIVISION

DAPHNIS

PROMETHEUS

JANUS

KELLER GAP

BEYOND

To E ring, inner large moons and Phoebe ring

The ring widths and distances are shown to scale, while the size of the moons are not.

RINGS – II

The striking bands of ice and rock that orbit Saturn.

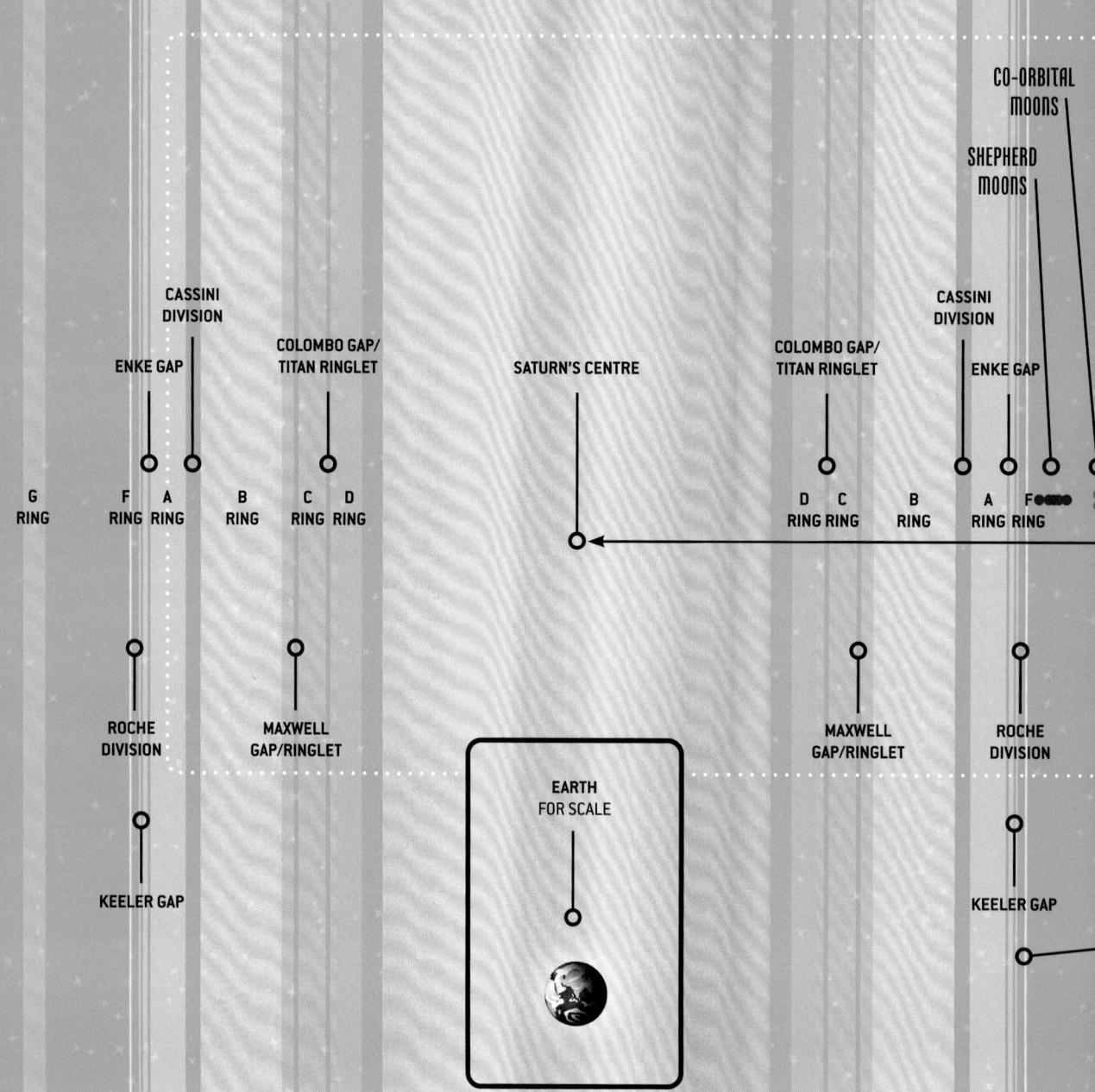

CO-ORBITAL
MOONS

SHEPHERD
MOONS

CASSINI
DIVISION

ENKE GAP

COLOMBO GAP/
TITAN RINGLET

SATURN'S CENTRE

COLOMBO GAP/
TITAN RINGLET

CASSINI
DIVISION

ENKE GAP

G
RING

F
RING

A
RING

B
RING

C
RING

D
RING

D
RING

C
RING

B
RING

A
RING

F
RING

ROCHE
DIVISION

MAXWELL
GAP/RINGLET

MAXWELL
GAP/RINGLET

ROCHE
DIVISION

EARTH
FOR SCALE

KEELER GAP

KEELER GAP

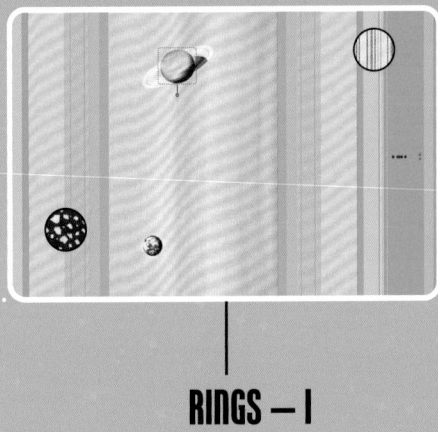

RINGS — I

See previous spread

MIMAS   ENCELADUS   DIONE   TETHYS

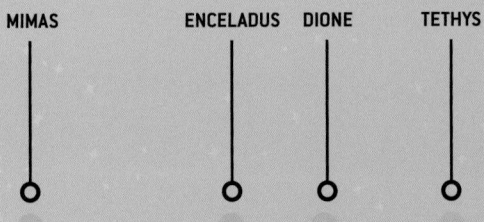

E RING
180,000 – 480,000 km from Saturn's centre

OUTER MOONS

To Rhea, Titan*,
Hyperion and Iapetus

*Titan is Saturn's
largest moon and the
second largest in the
Solar System. See
SEAS OF TITAN 144

OUTER RINGS

Dark outer rings detected by space-
based probes and telescopes.

The ring widths and distances are shown to
scale, while the size of the moons are not.

RINGS – III

The striking bands of ice and rock that orbit Saturn.

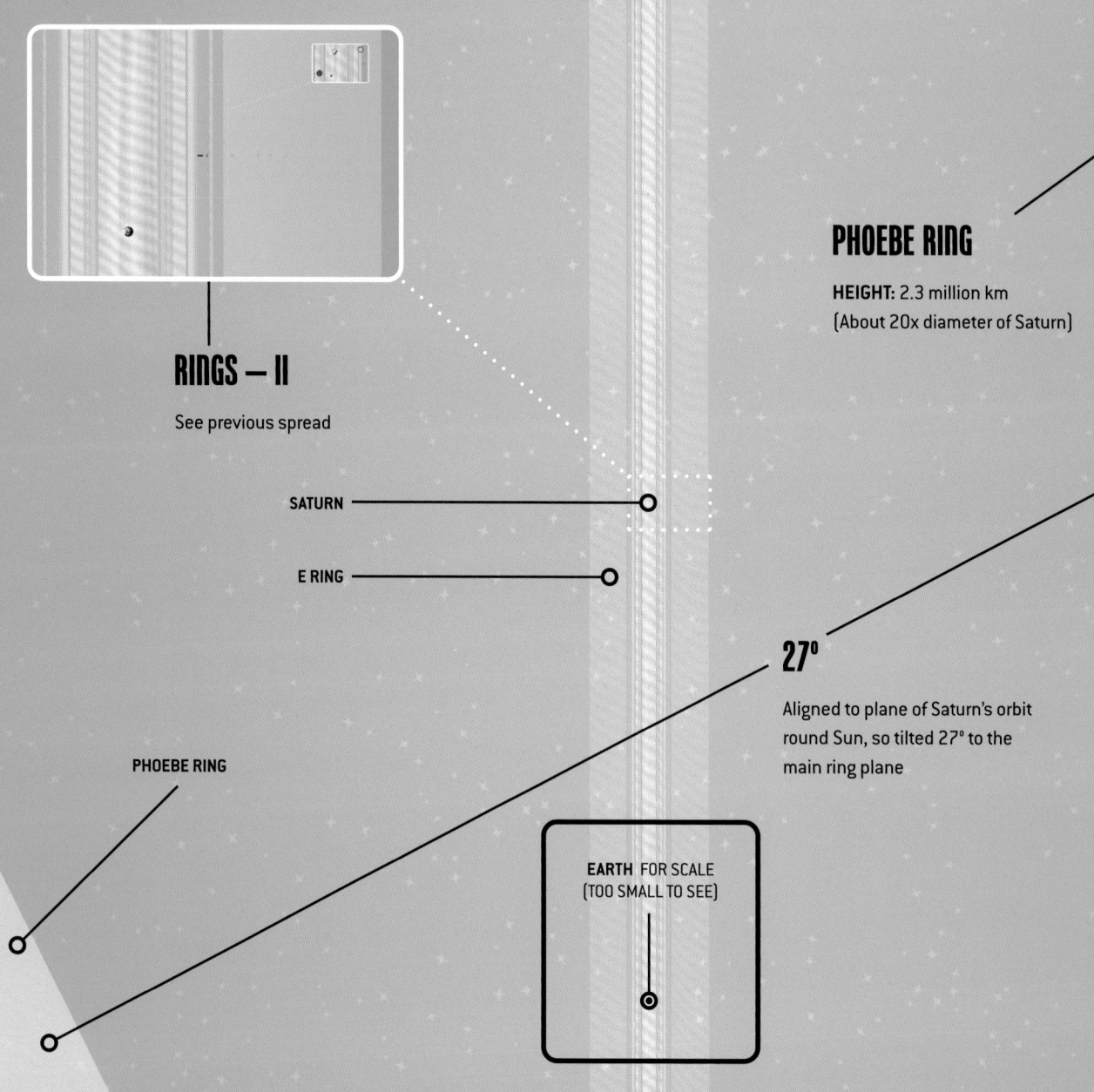

RINGS – II

See previous spread

PHOEBE RING

HEIGHT: 2.3 million km
(About 20x diameter of Saturn)

SATURN

E RING

27°

Aligned to plane of Saturn's orbit round Sun, so tilted 27° to the main ring plane

PHOEBE RING

EARTH FOR SCALE
(TOO SMALL TO SEE)

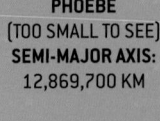

PHOEBE
(TOO SMALL TO SEE)
SEMI-MAJOR AXIS:
12,869,700 KM

ICE AND DUST

Made up of ice and dust particles
only visible in infrared

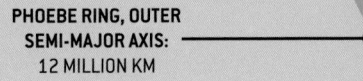

PHOEBE RING, INNER
SEMI-MAJOR AXIS:
6 MILLION KM

PHOEBE RING, OUTER
SEMI-MAJOR AXIS:
12 MILLION KM

HIDDEN RINGS

Obliquity means the rings change orientation to Earth
over a 30-year cycle — and every 15 years the rings are
edge-on to Earth and almost invisible.

Saturn appears brightest when the rings are face-on
to us, and dimmest when edge-on.

MOON TYPES – I

Saturn has 146 moons — more than any other planet in the Solar System!

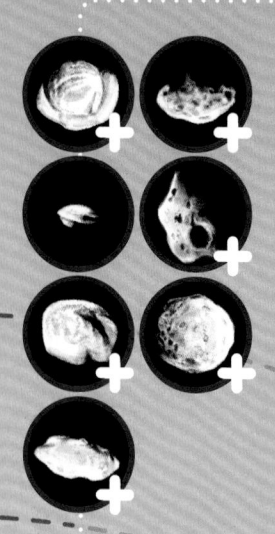

5X RING SHEPHERDS & 2X CO-ORBITAL MOONS

PAN+, DAPHNIS, ATLAS+, PROMETHEUS+, PANDORA+, EPIMETHEUS+ AND JANUS+

Small, irregularly shaped moons within or near the rings. Their gravitational force creates the gaps between the rings and also the rings' "sharp" edges.

Janus and Epimetheus are almost the same distance from Saturn. Instead of colliding, gravitational forces mean they swap orbits every four Earth years.

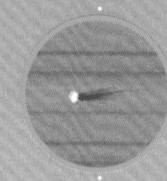

2X RING MOONLETS

INCLUDING S/2009 S1, AEGAEON AND 100+ VERY SMALL MOONLETS

Tiny (40–500 km in diameter) moonlets within the rings and not big enough to clear a continuous gap. Often so small we can't see them directly, only the shadows they cast.

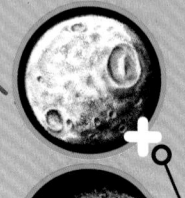

4X INNER LARGE MOONS

MIMAS+, ENCELADUS+, TETHYS+, DIONE+

Big enough to be round, these moons sit within the E ring.

Mimas: Prominent impact crater one-third of its diameter.

Enceladus:

* "Tiger stripe" fractures
* Cryovolcanism = jets of water vapour and dust
* Creates Saturn's E ring

Tethys: Composed mainly of water ice rather than rock.

Dione: Shows evidence of past tectonic activity.

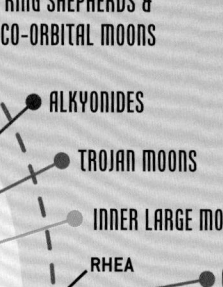

E RING

SATURN   INNERMOST
MOON: S/2009 S1

RING MOONLETS

RING SHEPHERDS &
CO-ORBITAL MOONS

ALKYONIDES

TROJAN MOONS

INNER LARGE MOONS

RHEA

OUTER LARGE MOONS

LARGEST MOON:
TITAN   HYPERION

KEY

 Image of moon not available

DIAMETER: ✚ More than 10 km   ✚✚ More than 3,000 km

3X ALKYONIDES

METHONE, ANTHE, PALLENE

Tiny moons that create "ring arcs" along their orbital paths; Pallene has a complete ring.

Methone: Egg-shaped and very smooth surface.

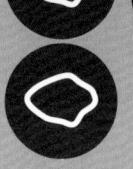

4X TROJAN MOONS

TELESTO+ AND CALYPSO+ (in same orbital path as Tethys), HELENE+ AND POLYDEUCES (in same orbital path as Dione).

Unique to Saturn, trojan moons share the orbit of another larger moon. Tethys and Dione both have two trojan moons each — one just ahead and one just behind as they orbit Saturn.

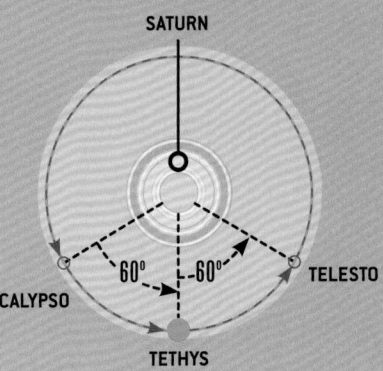

SATURN

CALYPSO   60° — 60°   TELESTO

TETHYS

4X OUTER LARGE MOONS

RHEA+, TITAN++, HYPERION+, IAPETUS+

Includes Saturn's three largest moons.

Rhea: Saturn's second-largest moon

Titan: Saturn's largest moon is the second largest moon in the Solar System. Lakes of liquid methane-ethane and subsurface ocean that erupts in cryovolcanoes.

Hyperion: Irregular shape and locked in orbit with Titan — it completes three orbits for every four orbits by Titan.

Iapetus: Saturn's third-largest moon with distinctive two-tone surface.

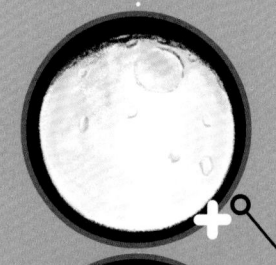

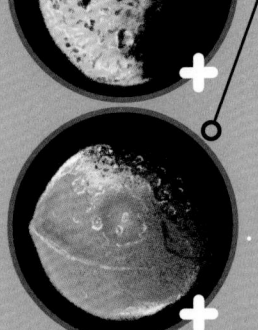

Distances are shown to scale, while the size of Saturn and the moons are not.

IAPETUS

MOON TYPES – II

12X INUIT GROUP

S/2019 S 1, KIVIUQ +, S/2005 S 4, S/2020 S 1, IJIRAQ +, PAALIAQ+, S/2004 S 31, TARQEQ, S/2019 S 14, SIARNAQ +, S/2020 S 3, S/2020 S 5

Small, irregular satellites of light-red colour, which may be remnants of one larger object.

?

2X NORSE OR GALLIC?

S/2004 S 24 and S/2006 S 12 may be either Inuit or Gallic moons.

7X GALLIC GROUP

ALBIORIX +, BEBHIONN, S/2007 S 8, S/2004 S 29, ERRIAPUS +, TARVOS+, S/2020 S 4

Small, irregular satellites of (different) light-red colour.

?

Distances are shown to scale, while the size of Saturn and the moons are not.

SATURN

LARGEST MOON:
TITAN

RHEA   HYPERION   IAPETUS

KIVIUQ

PHOEBE

100X NORSE GROUP

INCLUDING PHOEBE +, SKATHI, SKOLL, HYRROKKIN, GREIP, MUNDILFARI, GRIDR, BERGELMIR, JARNSAXA, NARVI, SUTTUNGR, HATI, EGGTHER, FARBAUTI, THRYMR, BESTLA, ANGRBODA, AEGIR, BELI, GERD, GUNNLOD, SKRYMIR, ALVALDI, KARI, GEIRROD, FENRIR, SURTUR, LOGE, YMIR +, THIAZZI, FORNJOT AND 69 AS-YET UNNAMED MOONS

Small, distant retrograde moons – they orbit in the opposite direction to the other moons and Saturn's own spin. This is probably due to having been captured by Saturn's gravity.

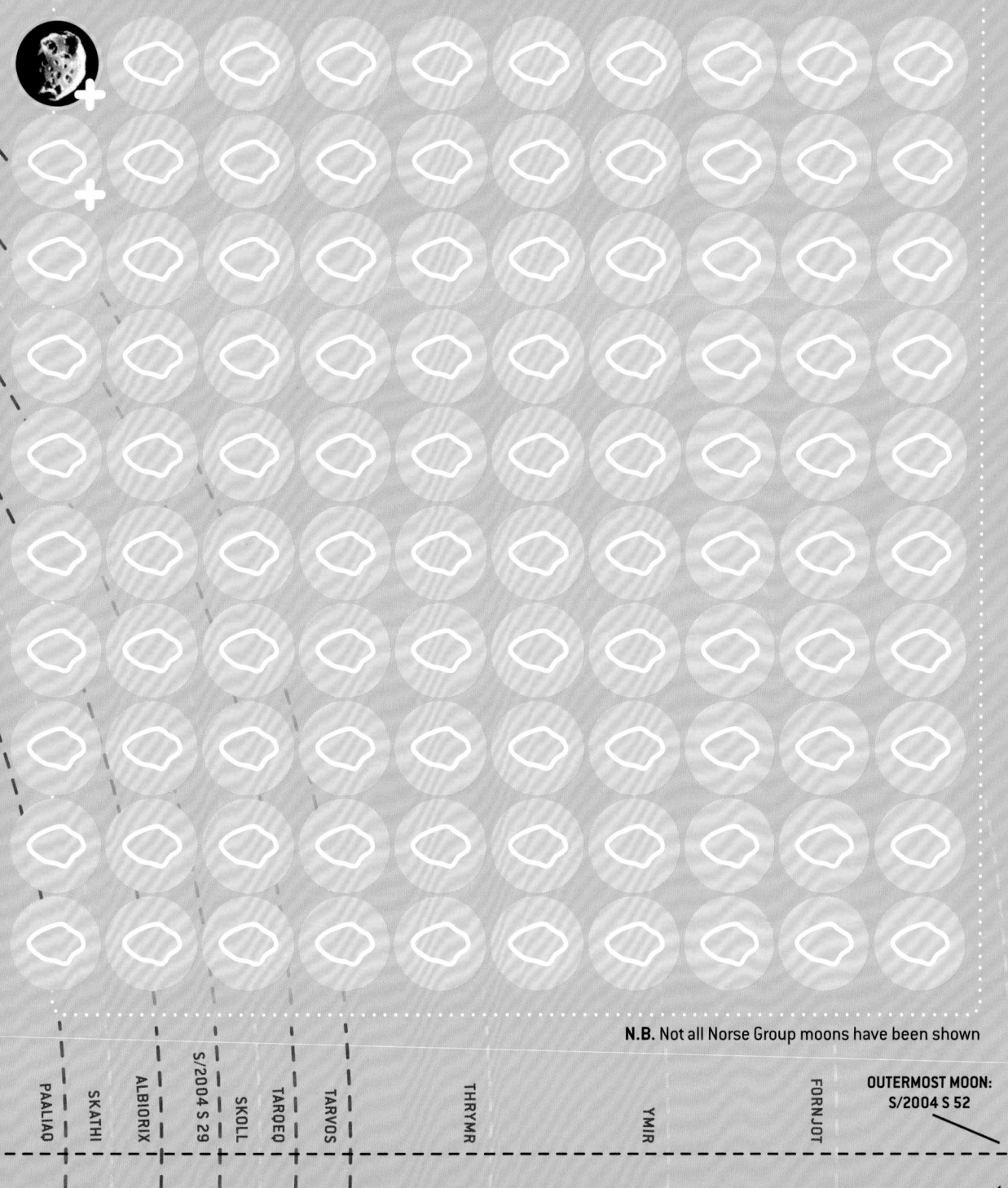

N.B. Not all Norse Group moons have been shown

PAALIAQ | SKATHI | ALBIORIX | S/2004 S 29 | SKOLL | TARQEQ | TARVOS | THRYMR | YMIR | FORNJOT | **OUTERMOST MOON:** **S/2004 S 52**

SEAS OF TITAN

Seas and lakes of methane and ethane on Saturn's moon Titan.

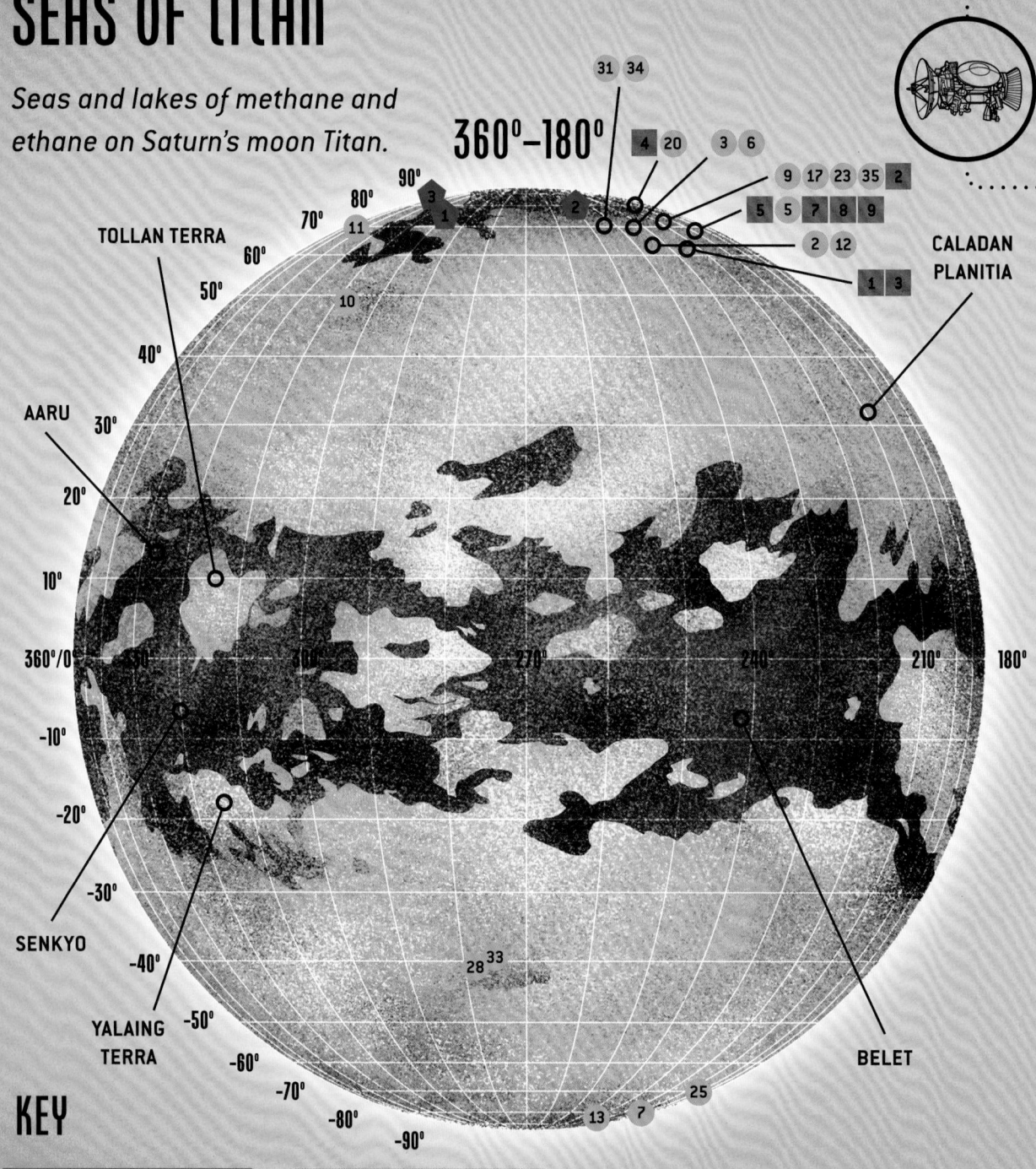

360°–180°

TOLLAN TERRA

AARU

SENKYO

YALAING TERRA

CALADAN PLANITIA

BELET

KEY

Large Seas ("Maria")

1. Kraken – longest dimension
1,170 km, area 400,000 km²

2. Ligeia – 500 km, area
126,000 km²

3. Punga – 380 km

Small Lakes ("Lacūs")

1. Abaya – 65 km

2. Albano – 6.2 km

3. Atitlán – 13.7 km

4. Bolsena – 101 km

5. Cardiel – 22 km

6. Cayuga – 22.7 km

7. Crveno – 41 km

8. Feia – 47 km

9. Freeman – 26 km

10. Hammar – 200 km

11. Jingpo – 240 km

12. Junin – 6.3 km

13. Kayangan – 6.2 km

14. Kivu – 77.5 km

15. Koitere – 68 km

16. Ladoga – 110 km

17. Lanao – 34.5 km

18. Logtak – 14.3 km

19. Mackay – 180 km

20. Müggel – 170 km

21. Mývatn – 55 km

22. Neagh – 98 km

23. Ohrid – 17.3 km

24. Oneida – 51 km

25. Ontario – 235
km and shallow, with
maximum depth of 7.4 m

26. Sevan – 46.9 km

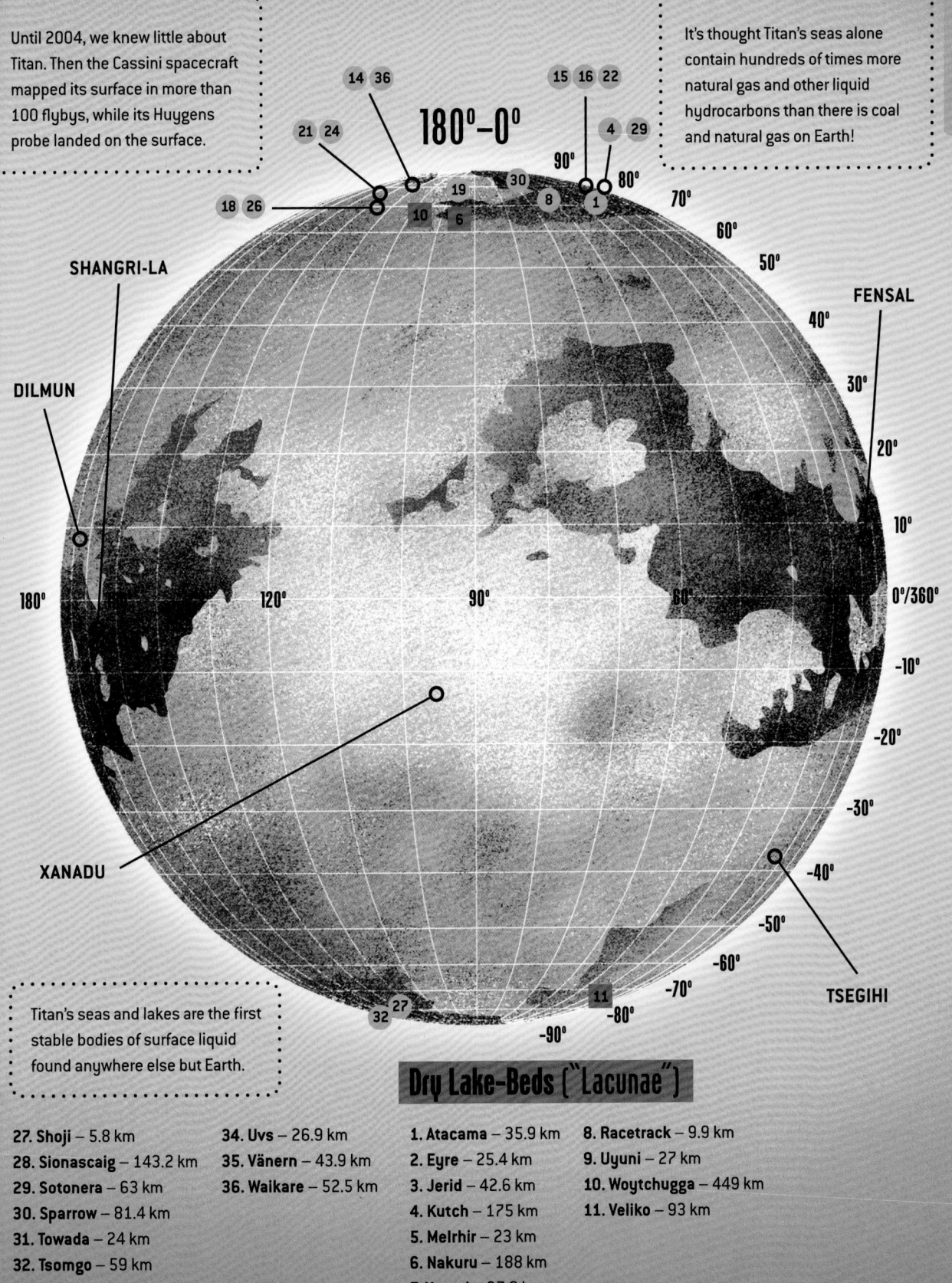

Until 2004, we knew little about Titan. Then the Cassini spacecraft mapped its surface in more than 100 flybys, while its Huygens probe landed on the surface.

It's thought Titan's seas alone contain hundreds of times more natural gas and other liquid hydrocarbons than there is coal and natural gas on Earth!

14 36

15 16 22

21 24

4 29

180⁰–0⁰

90⁰

19

80⁰

18 26

30

70⁰

10   6

8

1

60⁰

50⁰

SHANGRI-LA

40⁰

FENSAL

30⁰

20⁰

DILMUN

10⁰

180⁰

120⁰

90⁰

60⁰

0⁰/360⁰

-10⁰

-20⁰

-30⁰

-40⁰

XANADU

-50⁰

-60⁰

TSEGIHI

-70⁰

11

Titan's seas and lakes are the first stable bodies of surface liquid found anywhere else but Earth.

32 27

-80⁰

-90⁰

Dry Lake-Beds ("Lacunae")

27. **Shoji** – 5.8 km
28. **Sionascaig** – 143.2 km
29. **Sotonera** – 63 km
30. **Sparrow** – 81.4 km
31. **Towada** – 24 km
32. **Tsomgo** – 59 km
33. **Urmia** – 28.6 km

34. **Uvs** – 26.9 km
35. **Vänern** – 43.9 km
36. **Waikare** – 52.5 km

1. **Atacama** – 35.9 km
2. **Eyre** – 25.4 km
3. **Jerid** – 42.6 km
4. **Kutch** – 175 km
5. **Melrhir** – 23 km
6. **Nakuru** – 188 km
7. **Ngami** – 37.2 km

8. **Racetrack** – 9.9 km
9. **Uyuni** – 27 km
10. **Woytchugga** – 449 km
11. **Veliko** – 93 km

10. URANUS

UPPER ATMOSPHERE

* Hydrogen, helium, and methane with deep clouds of ammonia crystals
* Methane in upper atmosphere absorbs red light, hence blue-green colour
* At base of atmosphere, ammonia and water ions exist in fluid layer and generate magnetic field

LOWER ATMOSPHERE

* From about 10,000 km down, ices of water, ammonia and methane (Uranus and Neptune are often called "ice giants"

BULK DENSITY = 1,600 KG/M³

MANTLE

* Hot, thick ice
* 5,111 km thick

SURFACE

* Minimum surface temperature −224°C — the coldest planet in the Solar System
* Featureless blue-green disc with few clouds, but methane haze in upper atmosphere hides storms in the cloud deck below

CORE

* Thought to contain small, solid silicate core
* 14,000 km diameter
* No internal heat source, unlike the other gas giants

RINGS

* Very dark, faint series of rings, made up of small particles up to a few cm across
* Temperature of rings = −195°C, much warmer than the surrounding, empty space (about −273°C), suggesting a heat source we don't yet understand

KEY FACTS

Diameter: 52,400 km
Average distance from Sun: 2,900 million km
Average distance from Earth: 2,724 million km
Day: 17 hours
Year: 84 Earth years
Moons: 28

URANUS

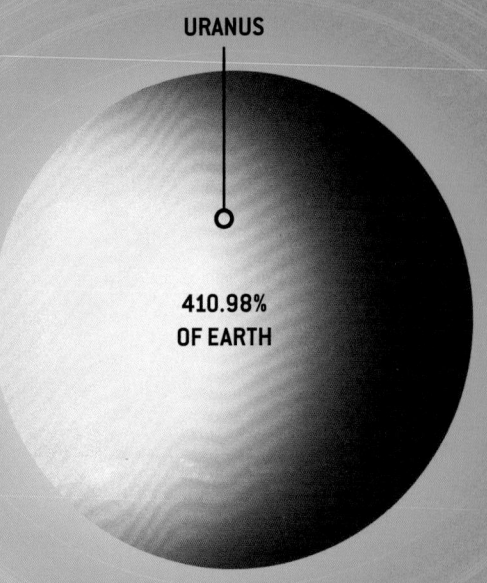

410.98%
OF EARTH

SCALE

EARTH

ROLL OVER

Uranus has an axial tilt (obliquity) of 98° — so rotates on its side compared to other planets.

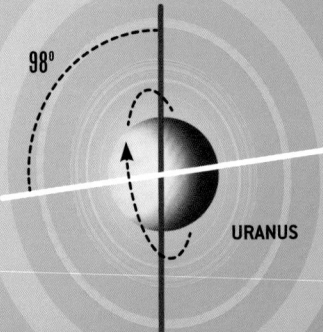

ROTATIONAL AXIS

23.4°

98°

EARTH

URANUS

SMELLY PLANET

The cloud tops of Uranus contain hydrogen sulphide — the same foul-smelling gas in rotten eggs.

URANUS

URANIAN MOONS

The least massive system of moons of the four outer planets.

15. ARIEL
(191,020 KM)

18. OBERON
(583,520 KM)

5X LARGE MOONS

MIRANDA, ARIEL, UMBRIEL, TITANIA AND OBERON

Large, dark and round moons.

* **Titania** and **Oberon** discovered 1787; **Ariel** and **Umbriel** 1851; **Miranda** 1948

* Heavily cratered, little atmosphere

* **Miranda** smallest, about 472 km diameter; **Titania** largest, about 1577 km diameter (about half the size of Earth's Moon)

16. UMBRIEL
(266,300 KM)

14. MIRANDA
(129,390 KM)

13X INNER MOONS

INCLUDING CORDELIA AND OPHELIA

Small, dark and irregular moons mostly composed of water ice.

* Range from about 18 km to about 162 km in diameter

* Chaotic, unstable orbital system, with much perturbation (affect on each other's orbits), possibly resulting in crossed orbits and collisions

* They are related to the Uranian rings. The two innermost moons, Cordelia and Ophelia, are shepherd moons for the ε ring. Mab is the source of Uranus's outermost μ ring.

* 11 first seen by Voyager 2 in 1986, Cupid and Mab by telescope in 2003

19. FRANCISCO
(4,276,000 KM)

URANUS

17. LARGEST MOON: TITANIA
(435,910 KM)

21. STEPHANO
(8,002,000 KM)

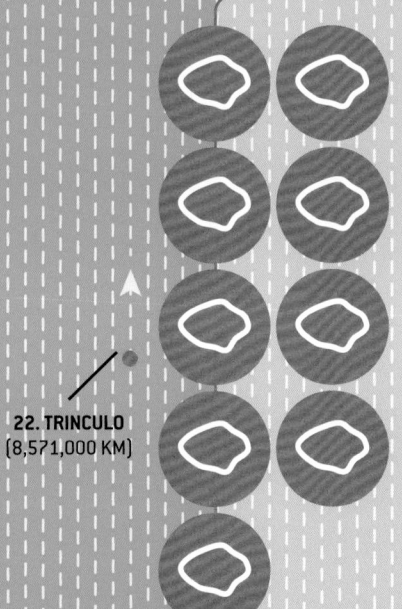

9X IRREGULAR MOONS

INCLUDING CALIBAN AND SYCORAX

Small, dark and irregular moons —
all probably objects captured by
Uranian gravity.

✳ Discovered between 1997 and
   2003

✳ Range from about 18 km to
   about 165 km in diameter

✳ All but **Margaret** = retrograde
   (their orbital path is in the
   opposite direction to the spin
   of Uranus)

22. TRINCULO
(8,571,000 KM)

25. PROSPERO
(16,418,000 KM)

27. FERDINAND
(20,900,000 KM)

24. MARGARET
(14,345,000 KM)

26. SETEBOS
(17,459,000 KM)

23. SYCORAX
(12,179,000 KM)

20. CALIBAN
(7,230,000 KM)

200,000 Km

PLANETARY MOTION

Kepler's three laws of planetary motion explain the elliptical paths of the planets.

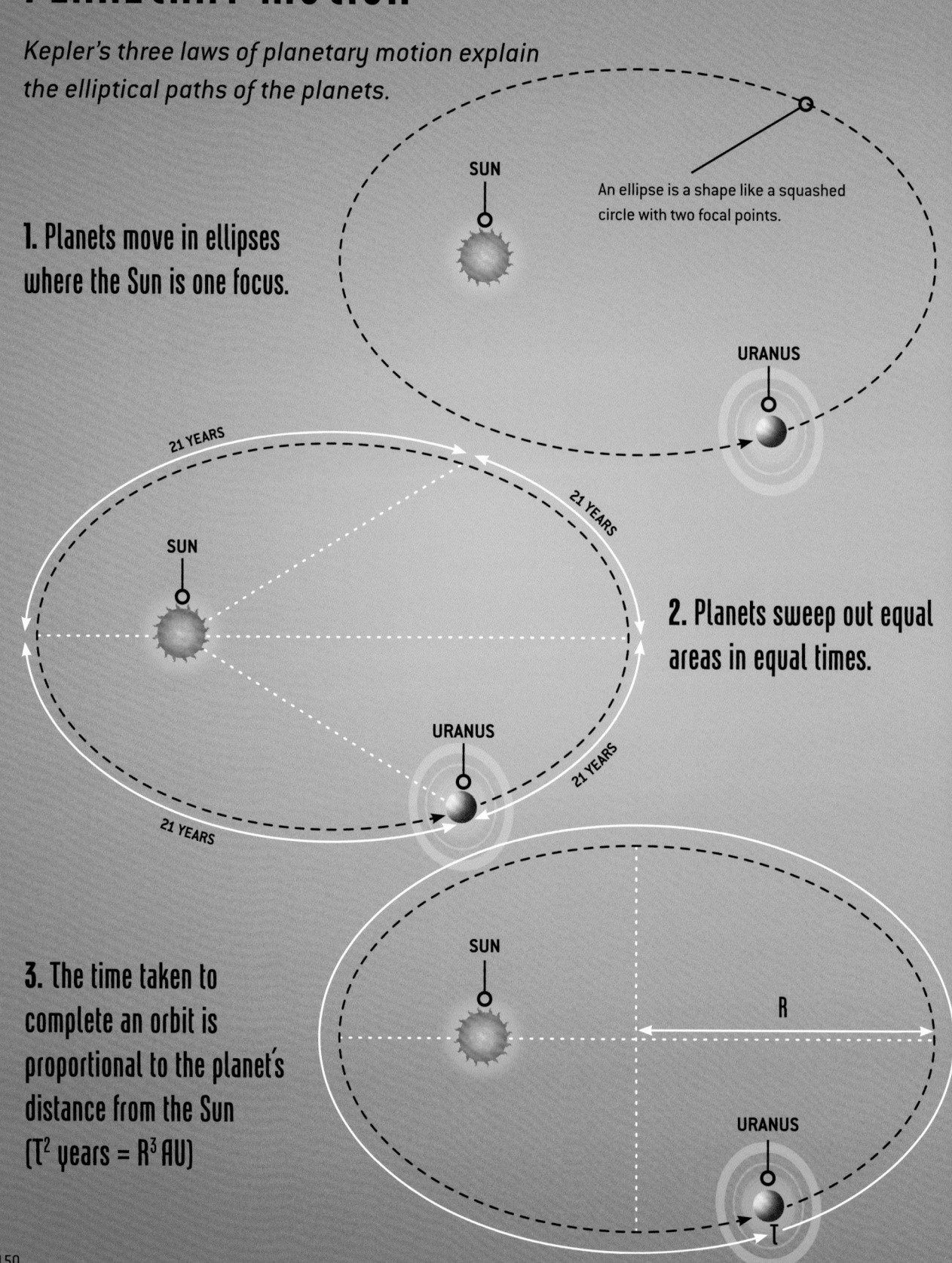

1. Planets move in ellipses where the Sun is one focus.

An ellipse is a shape like a squashed circle with two focal points.

SUN

URANUS

21 YEARS

21 YEARS

21 YEARS

21 YEARS

SUN

URANUS

2. Planets sweep out equal areas in equal times.

3. The time taken to complete an orbit is proportional to the planet's distance from the Sun (T^2 years = R^3 AU)

SUN

R

URANUS

KEPLER'S THREE LAWS, WITH TWO PLANETARY ORBITS

The orbits are ellipses, with focal points **F1** and **F2** for the first planet and **F1** and **F3** for the second planet. The Sun is placed in focal point **F1**.

The two shaded sectors **A1** and **A2** have the same surface area and the time for **planet 1** to cover segment **A1** is equal to the time to cover segment **A2**.

The total orbit times for **planet 1** and **planet 2** have a ratio **a1:a2 = 3:2**

OTHER SYSTEMS

These planetary laws seem to be universal and apply to other systems such as the Moon orbiting Earth. We've also discovered planets orbiting other stars that follow these laws.

Measuring the time taken for these planets to complete an orbit, we can then work out how close they are to their star!

See **BEYOND > EXOPLANETS**

F1
SUN

a1

F2

A1

A2

PLANET 1
EARTH

a2

F3

PLANET 2
URANUS

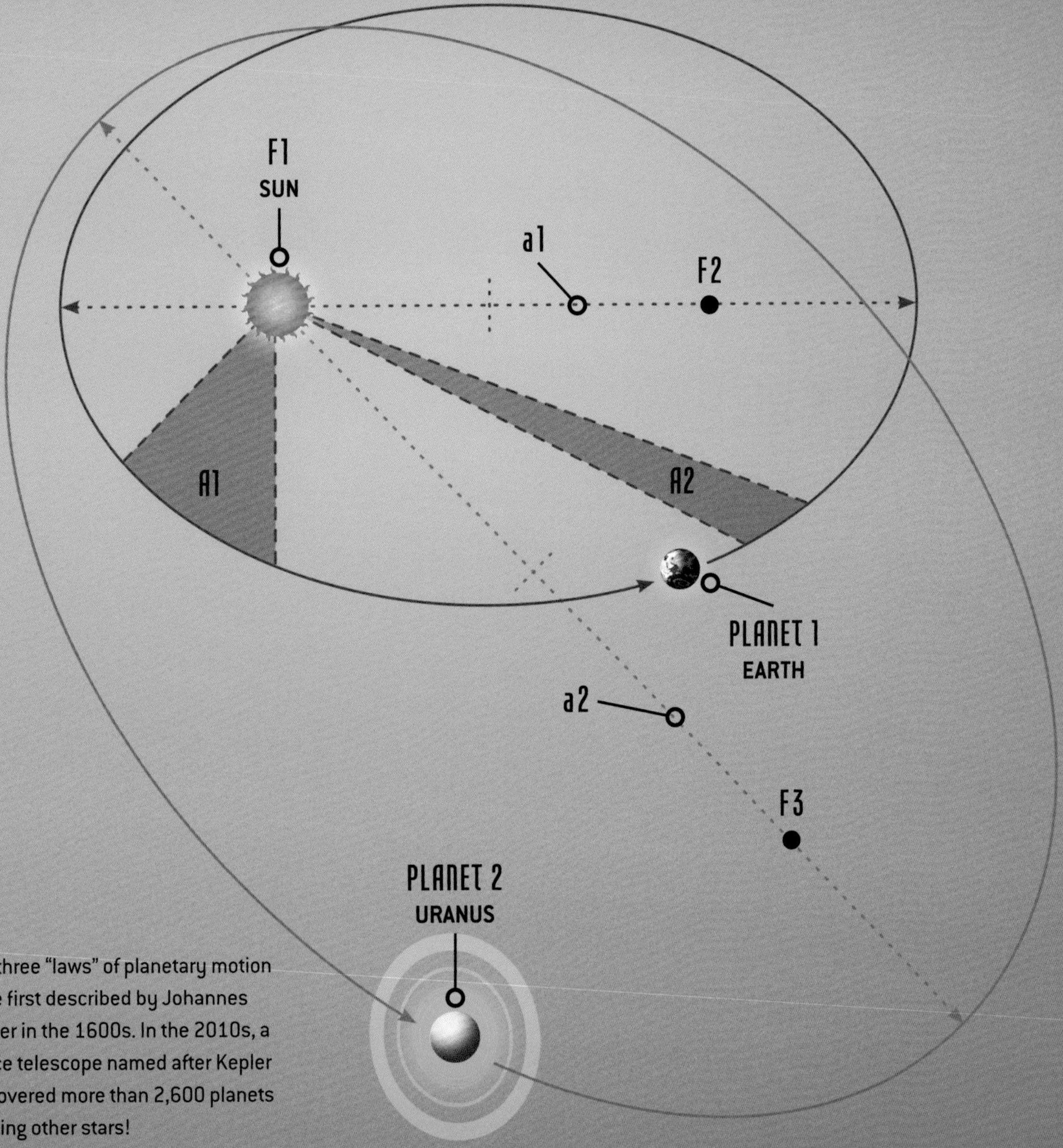

The three "laws" of planetary motion were first described by Johannes Kepler in the 1600s. In the 2010s, a space telescope named after Kepler discovered more than 2,600 planets orbiting other stars!

11. NEPTUNE

INNER ATMOSPHERE

 * About 80% of Neptune's mass = hot, dense fluid of water, methane and ammonia ice

ATMOSPHERE

 * Atmosphere of hydrogen, helium and methane gas

SURFACE

 * Average surface temperature = −214°C

 * Winds reach 2,000 km an hour — the fastest in the Solar System

CORE

 * Small, probably rocky core

MASS AND DENSITY

 * Slightly smaller than Uranus but has higher mass

 * Bulk density is the same as Uranus (1,600 kg/m³), so probably has similar internal structure

COLOUR

 * Absorption of red light by methane = blue colour (we don't yet know why it's a more vivid blue than Uranus!)

STORMS

 * Large, short-lived storms seen as bright and dark spots in upper atmosphere

RINGS

 * Ring system of dark particles, with thicker clumps of material spread out along arcs in several rings

KEY FACTS

Diameter: 49,500 km
Average distance from Sun: 4,500 million km
Average distance from Earth: 4,351 million km
Day: 16 hours
Year: 165 Earth years
Moons: 16

SCALE

NEPTUNE

388.24%
OF EARTH

EARTH

COLD BUT NOT THAT COLD

Uranus has a lower minimum temperature than
Neptune, but Neptune is colder on average. Yet it's
still warmer than we would predict given its distance
from the Sun, suggesting an internal heat source.

QUICK CHANGE

With the fastest planetary winds in the Solar System,
atmospheric features change rapidly on Neptune.
The Great Dark Spot (GDS) storm seen by Voyager 2 in
1989 had disappeared by the time of Hubble images
in 1994. In 2018, Hubble images showed a new
storm — Dark Spot 2!

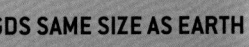

GDS SAME SIZE AS EARTH

NEPTUNE, 1989

NEPTUNE, 1994

GRAVITY AND THE DISCOVERY OF NEPTUNE

Gravitational force becomes less intense with distance, which helped us discover the planet Neptune.

All bodies with mass or energy **GRAVITATE** or move toward one another.

Gravity increases:

* The more massive the body
* The closer you are to that body

The latter is because the same gravitational attraction 2x as far away must cover an area 4x as big, so is 4x less strong.

This **INVERSE SQUARE LAW** also applies to electrical fields, light, sound and radiation.

The inverse square law helped astronomers to find the planet Neptune.

The observed position of Uranus did not quite match the orbit calculated by the three laws of **PLANETARY MOTION**. Sometimes it was a little ahead, sometimes it was a little behind.

See **URANUS > PLANETARY MOTION**  **150**

NEPTUNE

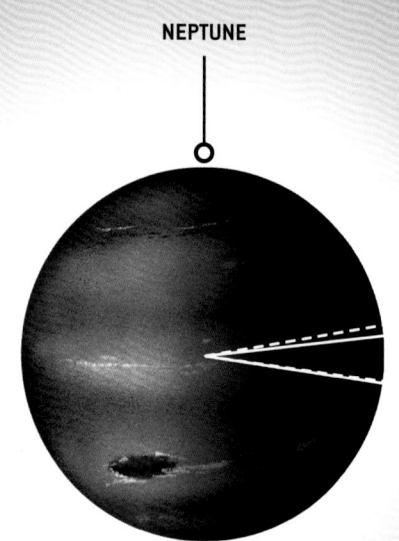

PREDICTED PATH AND ACTUAL PATH OF URANUS

By the 1830s, astronomers suggested its orbit was being perturbed by another, more distant planet — one not yet discovered.

When behind Uranus as they both orbited the Sun, the pull of this planet's gravity would slow Uranus.

When ahead of Uranus, the pull of this planet's gravity would make Uranus move faster.

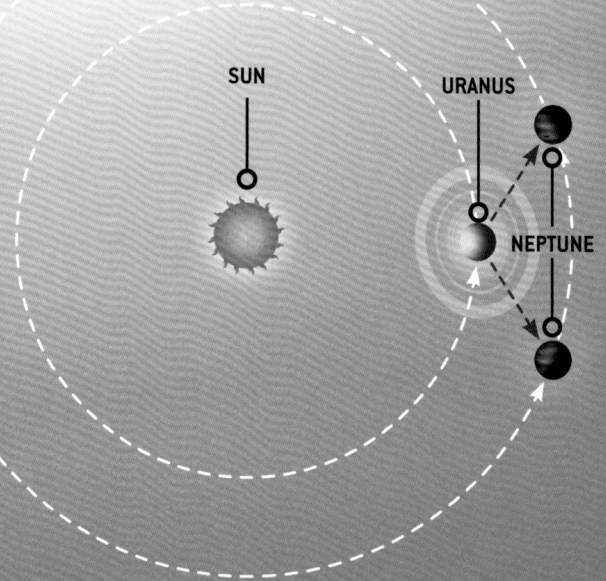

SUN

URANUS

NEPTUNE

KEY

- - - → **PULL OF GRAVITY**

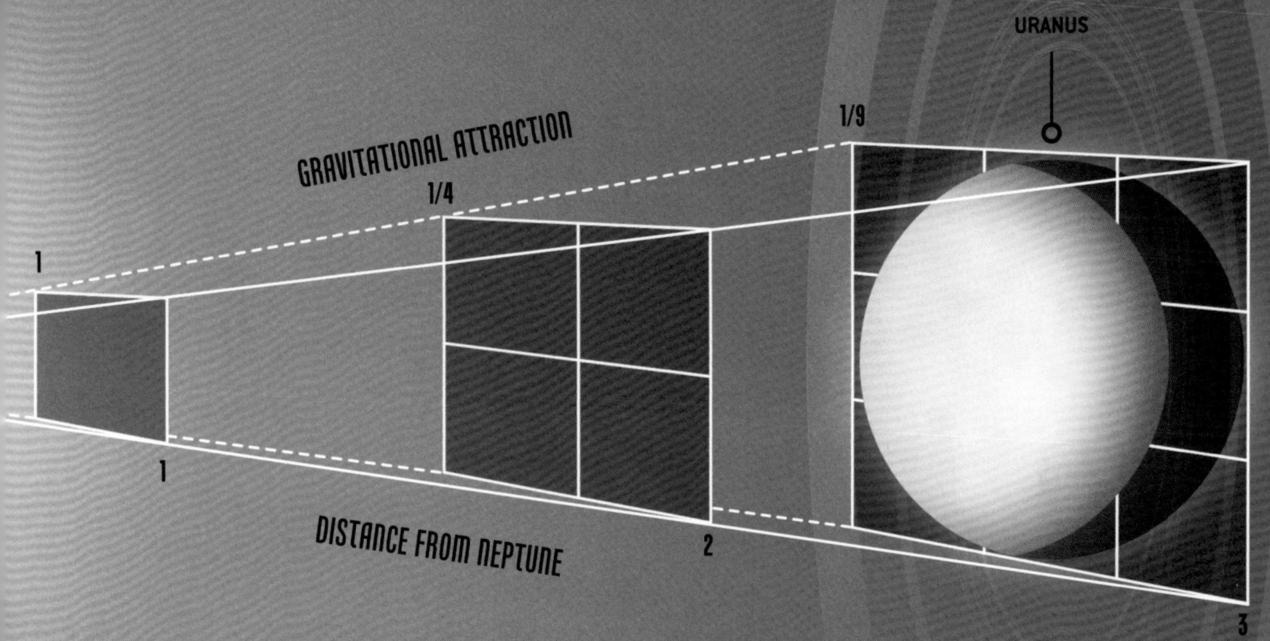

GRAVITATIONAL ATTRACTION

1/4

1/9

1

1

2

3

DISTANCE FROM NEPTUNE

URANUS

NOT TO SCALE

Using the inverse square law, Urbain Jean Joseph Le Verrier (1811-87) and John Couch Adams (1819-92) each calculated the position of a planet whose gravity would affect the orbit of Uranus in the way observed.

On 23 September 1846, the Berlin Observatory used Le Verrier's calculation to find Neptune — very close to where he said it would be.

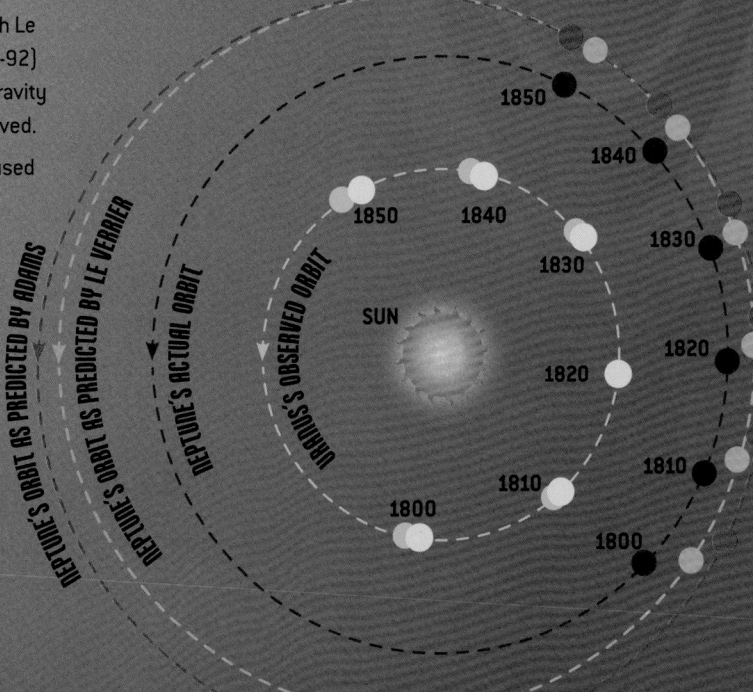

NEPTUNE'S ORBIT AS PREDICTED BY ADAMS

NEPTUNE'S ORBIT AS PREDICTED BY LE VERRIER

NEPTUNE'S ACTUAL ORBIT

URANUS'S OBSERVED ORBIT

SUN

1850
1840
1830
1820
1810
1800

1850
1840
1830
1820
1810
1800

KEY

- URANUS'S PREDICTED POSITION IF NEPTUNE DID NOT EXIST
- URANUS'S OBSERVED POSITION
- NEPTUNE'S POSITION AS PREDICTED BY LE VERRIER
- NEPTUNE'S POSITION AS PREDICTED BY ADAMS
- NEPTUNE'S ACTUAL POSITION

PROBES – II

The craft we've sent to fly past or orbit worlds further from the Sun than we are.

KEY

STILL ACTIVE [AS OF JAN 2024]

CRASHED

FAILED

ASTEROID BELT – 16

1991 – Galileo [on way to Jupiter]

1997 – NEAR Shoemaker

1999 – Deep Space 1 [on way to comet]

2000 – Cassini [on way to Saturn]

2002 – Stardust [on way to comet]

2005 – Hayabusa [returned to Earth with samples 2010]

2007 – New Horizons [on way to Pluto]

2008 – Rosetta [on way to comet]

2011 – Dawn

2012 – Chang'e 2

2018 – Hayabusa2 [returned to Earth with samples 2020]

2018 – OSIRIS-REx [returned samples to Earth 2023]

2022 – DART

2023 – Lucy

2028 – JUICE [predicted, on way to Jupiter, launched 2023]

2029 – Psyche [predicted launched 2023]

GALILEO

2003 – crashed into Jupiter's atmosphere

JUPITER – 10

1973 – Pioneer 10

1974 – Pioneer 11 [on way to Saturn]

1979 – Voyager 1 [on way to Saturn]

1979 – Voyager 2 [on way to Saturn, Uranus, Neptune]

1992 – Ulysses

1995 – Galileo

2000 – Cassini [on way to Saturn]

2007 – New Horizons [on way to Pluto]

2016 – Juno

2031 – JUICE [predicted]

JUPITER, FAILED – 0

ASTEROID BELT, FAILED – 2

1991 – Clementine

2016 – PROCYON

MARS

See **MERCURY > PROBES – I**

30

URANUS — 1

- 1986 – Voyager 2 [on way to Neptune]

URANUS, FAILED — 0

NEPTUNE — 1

- 1989 – Voyager 2 [on way to interstellar space]

NEPTUNE, FAILED — 0

SATURN — 4

- 1979 – Pioneer 11
- 1980 – Voyager 1 [on way to interstellar space]
- 1981 – Voyager 2 [on way to Uranus, Neptune]
- 2004 – Cassini

SATURN, FAILED — 0

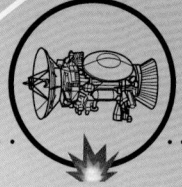

CASSINI-HUYGENS

2017 – crashed into Saturn

See **JUPITER > GRAVITATIONAL SLINGSHOT**

110

PLUTO — 1

- 2015 – New Horizons [on way to further Kuiper Belt Objects, still active]

PLUTO, FAILED — 0

COMETS — 10

* = No. of comets visited not including Halley's comet;
H = visited Halley's comet; **a** = visited asteroid

1985 – ICE: ***H**	2001 – Deep Space: **a***
1986 – Vega 1: **H**	2004 – Stardust / NExT: **a**** returned sample 2006
1986 – Suisei: **H**	
1986 – Vega 2: **H**	2005 – Deep Impact: *****
1986 – Sakigake (1986): **H**	2014 – Rosetta: **aa***
1986 – Giotto (1986): **H***	

COMETS, FAILED — 1

- 2003 – CONTOUR

BEYOND — 5

- Voyager 2
- New Horizons
- Pioneer 10
- Pioneer 11
- Voyager 1

See **SUN > A SENSE OF SCALE**

14

PLANETARY FORECAST – I

What the weather is like on the inner planets.

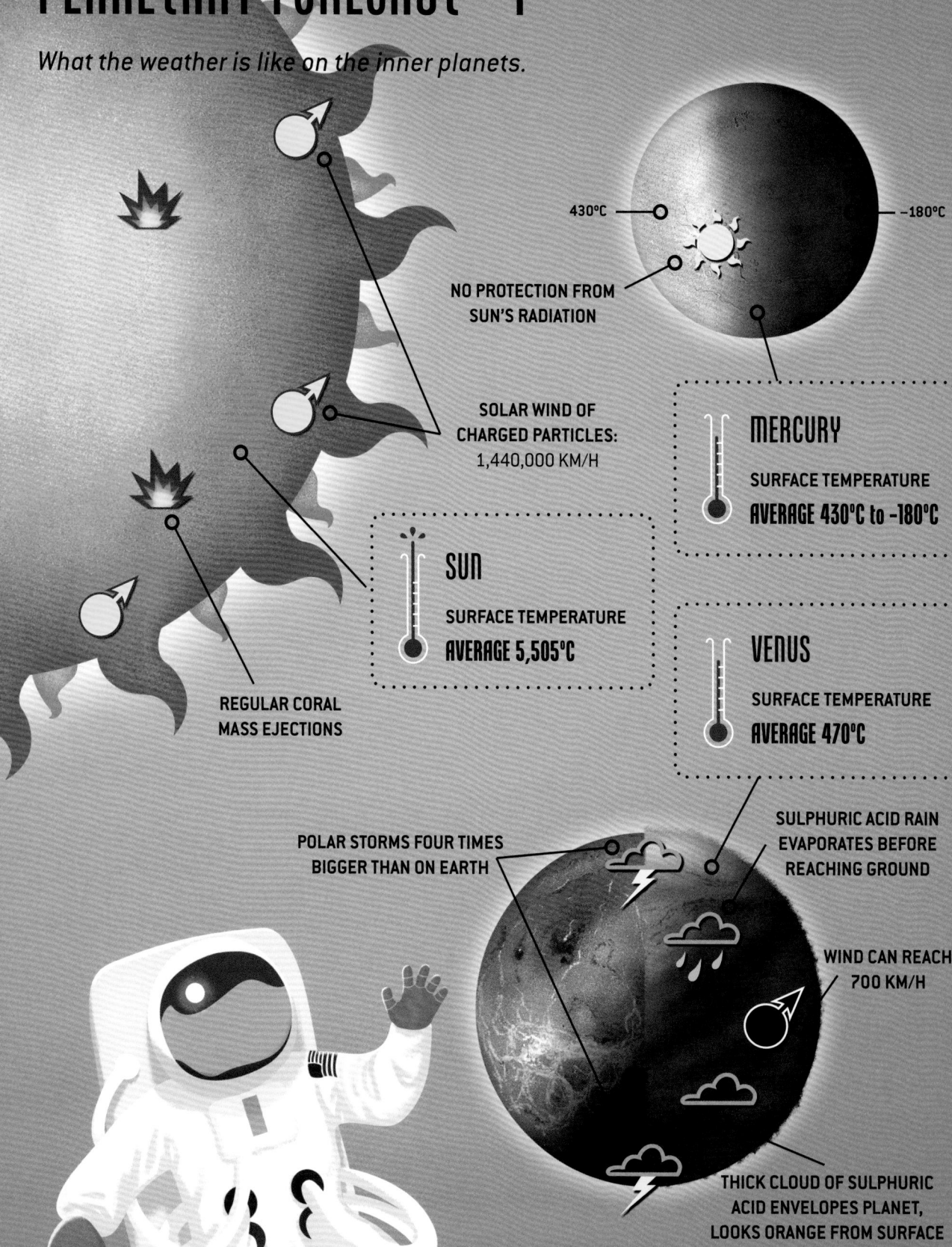

NO PROTECTION FROM
SUN'S RADIATION

430°C

−180°C

SOLAR WIND OF
CHARGED PARTICLES:
1,440,000 KM/H

MERCURY

SURFACE TEMPERATURE

AVERAGE 430°C to −180°C

SUN

SURFACE TEMPERATURE

AVERAGE 5,505°C

REGULAR CORAL
MASS EJECTIONS

VENUS

SURFACE TEMPERATURE

AVERAGE 470°C

POLAR STORMS FOUR TIMES
BIGGER THAN ON EARTH

SULPHURIC ACID RAIN
EVAPORATES BEFORE
REACHING GROUND

WIND CAN REACH
700 KM/H

THICK CLOUD OF SULPHURIC
ACID ENVELOPES PLANET,
LOOKS ORANGE FROM SURFACE

COMPLEX, VARIED WEATHER

EARTH

SURFACE TEMPERATURE

AVERAGE ~14°C (but much variation)

WATER RAIN

FASTED WIND
RECORDED: 372 KM/H

POLAR ICE CAPS GROW AND
SHRINK WITH SEASONS

FREQUENT SMALL
DUST STORMS

OCCASIONAL, HUGE DUST
STORMS ENCIRCLE PLANET!

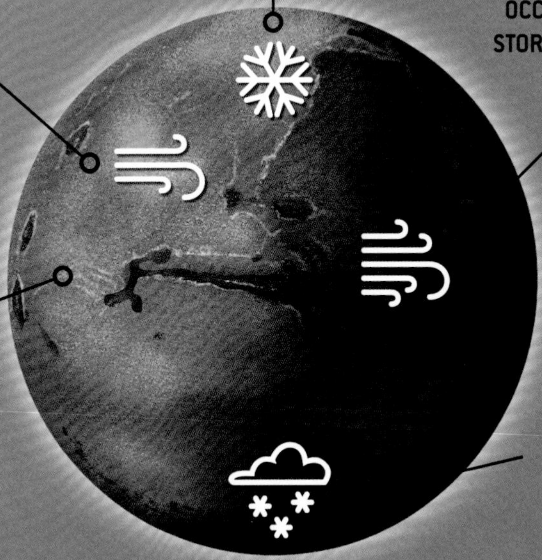

MARS

SURFACE TEMPERATURE

AVERAGE –60°C

CARBON DIOXIDE SNOW
(may fall from sky or in
the air at surface level)

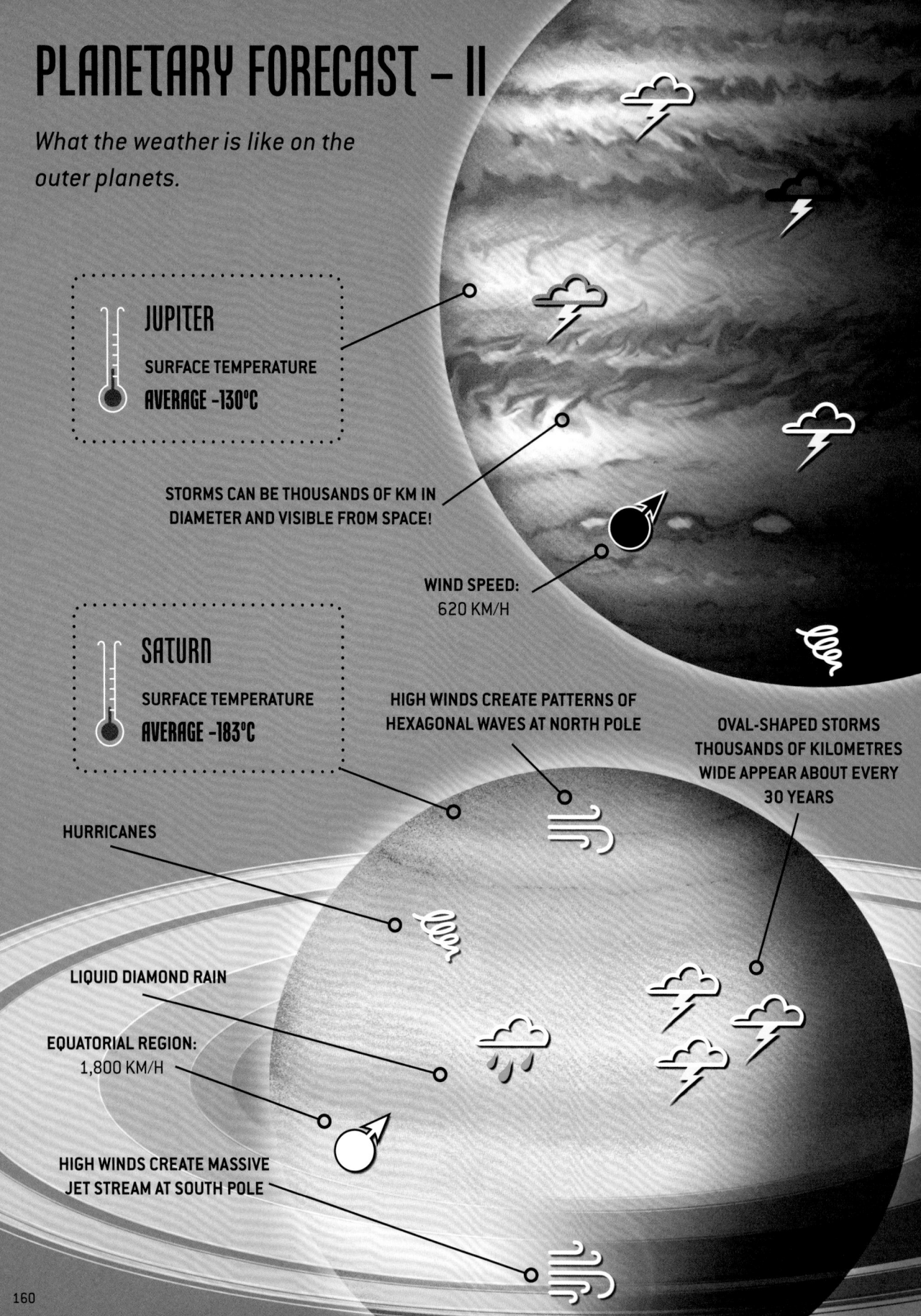

PLANETARY FORECAST – II

What the weather is like on the outer planets.

JUPITER

SURFACE TEMPERATURE
AVERAGE -130°C

STORMS CAN BE THOUSANDS OF KM IN
DIAMETER AND VISIBLE FROM SPACE!

WIND SPEED:
620 KM/H

SATURN

SURFACE TEMPERATURE
AVERAGE -183°C

HIGH WINDS CREATE PATTERNS OF
HEXAGONAL WAVES AT NORTH POLE

OVAL-SHAPED STORMS
THOUSANDS OF KILOMETRES
WIDE APPEAR ABOUT EVERY
30 YEARS

HURRICANES

LIQUID DIAMOND RAIN

EQUATORIAL REGION:
1,800 KM/H

HIGH WINDS CREATE MASSIVE
JET STREAM AT SOUTH POLE

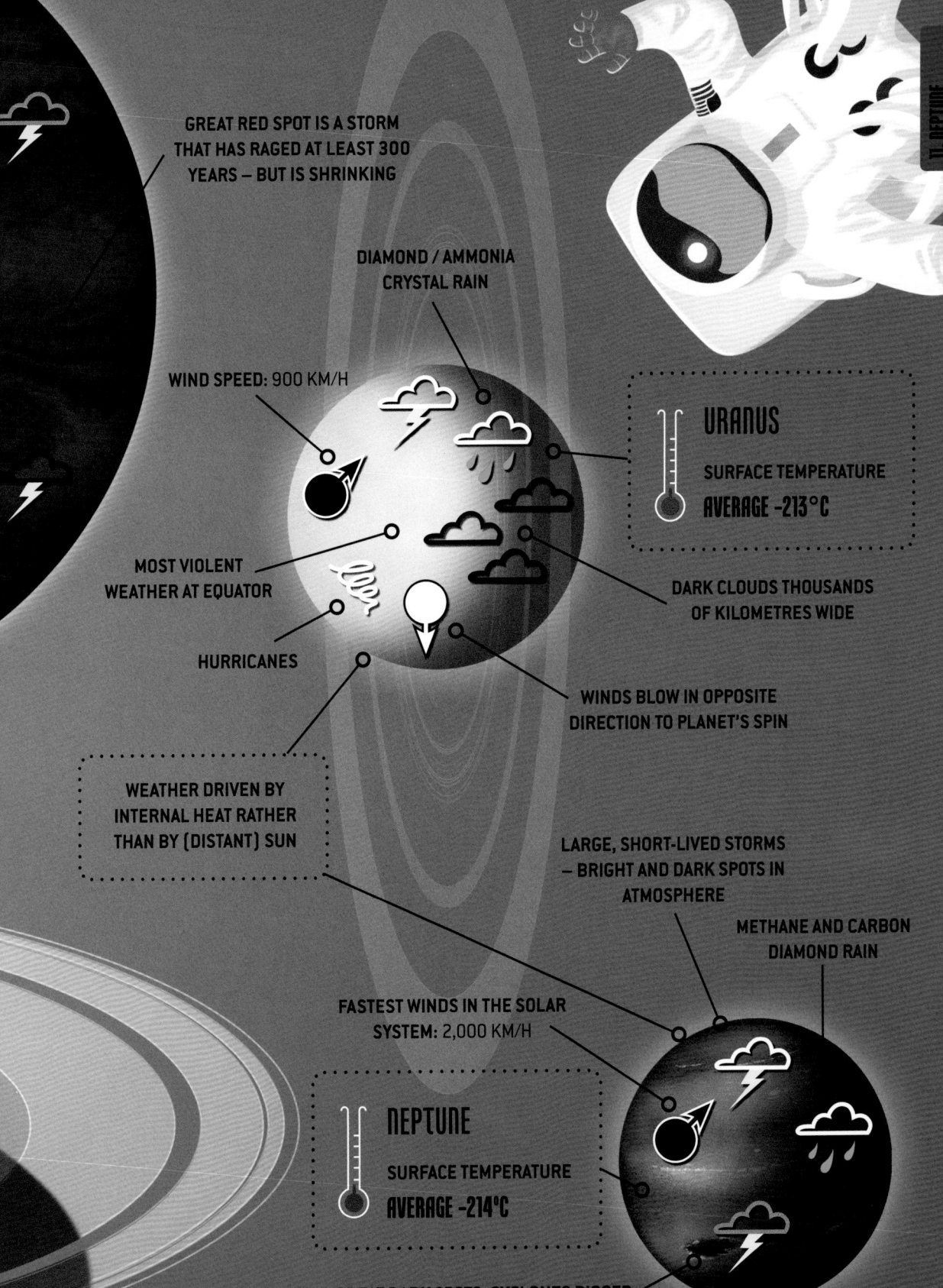

GREAT RED SPOT IS A STORM THAT HAS RAGED AT LEAST 300 YEARS – BUT IS SHRINKING

DIAMOND / AMMONIA CRYSTAL RAIN

WIND SPEED: 900 KM/H

URANUS
SURFACE TEMPERATURE
AVERAGE -213°C

MOST VIOLENT WEATHER AT EQUATOR

DARK CLOUDS THOUSANDS OF KILOMETRES WIDE

HURRICANES

WINDS BLOW IN OPPOSITE DIRECTION TO PLANET'S SPIN

WEATHER DRIVEN BY INTERNAL HEAT RATHER THAN BY (DISTANT) SUN

LARGE, SHORT-LIVED STORMS – BRIGHT AND DARK SPOTS IN ATMOSPHERE

METHANE AND CARBON DIAMOND RAIN

FASTEST WINDS IN THE SOLAR SYSTEM: 2,000 KM/H

NEPTUNE
SURFACE TEMPERATURE
AVERAGE -214°C

GREAT DARK SPOTS: CYCLONES BIGGER THAN EARTH WHICH COME AND GO

161

12. BEYOND

*Pluto is the best known of the objects in the Kuiper Belt, a "doughnut"
beyond Neptune and extending some 50 AU.*

ATMOSPHERE

* Thin atmosphere expands when
  Pluto's orbit brings it closer to the
  Sun and ices melt — just like
  with comets!

MANTLE

* Water ice

CORE-MANTLE BOUNDARY

* A liquid ocean layer
  of water and ammonia
  100–180 km thick

CORE

* Probably made of rock
  and ice

* 1,700 km in diameter
  — 70% of total!

SURFACE

* 98+% solid nitrogen, plus trace
  methane and carbon dioxide

* Temperature ranges from −226ºC to −240ºC

NEW HORIZONS

Surface features seen by New Horizons probe in 2015:

* Mountains of water ice, 2–3 km tall

* Valleys 600 km long

* Craters 260 km in diameter, with signs of erosion
  suggesting Pluto has plate tectonics

* Plains of frozen nitrogen gas

NO RINGS, NO FIELDS

* Pluto does not have rings

* No magnetic fields have been detected,
  and it is so small and rotates so slowly it
  probably doesn't have them

KEY FACTS

Diameter: 2,374 km
Average distance from Sun: 5,910 million km
Average distance from Earth: 5,900 million km
Day: 6 Earth days
Year: 248 Earth years
Moons: 5

SCALE

PLUTO

**18.62%
OF EARTH**

EARTH'S SIZE

EARTH

OVERLAPPING ORBIT

Pluto's orbit around the Sun is more **ECCENTRIC** — more egg-shaped instead of circular — than any planet in the Solar System.

11 FEBRUARY 1999

EARTH

NEPTUNE

7 FEBRUARY 1979

PLUTO

EX-PLANET

Pluto was a planet for 76 years, between its discovery in 1930 and the decision by the International Astronomical Union to formally define "planet" — and exclude it!
See **DWARF PLANETS** | 166 |

VISIONS OF PLUTO

Timeline of our knowledge of this distant world.

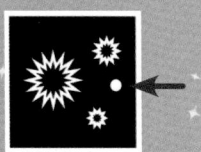

PLUTO   EARTH

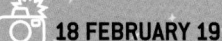

1930

📷 18 FEBRUARY 1930

After a year's search, 24 year-old Clyde Tombaugh spots a pinprick of light in different positions in two photographs of the same area of the night sky.

📷 13 MARCH 1930

Checking more photographs confirms that Tombaugh has discovered a new planet and the discovery is announced to the world.

1931

Mass of Pluto estimated to be 1x Earth based on its effect on orbits of Neptune and Uranus.

1948

Estimated mass of Pluto revised to 0.1x Earth, based on same information.

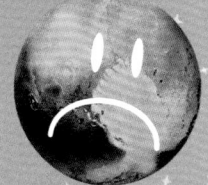

AUGUST 2006

With discovery of Eris and other such objects, the International Astronomical Union defines "planet" for the first time — and the definition excludes Pluto. Estimated mass of Pluto revised to 0.00218x Earth.

📷 FEBRUARY 2006

Photograph of Pluto and three moons by Hubble Space Telescope.

JANUARY 2006

New Horizons probe launched to visit Pluto.

📷 2005

Photograph of Pluto by Hubble Space Telescope.

MOONS: 4

CHARON
HYDRA
NIX
KERBEROS

MOONS: 5

CHARON
HYDRA
NIX
STYX
KERBEROS

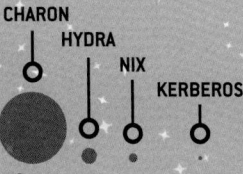

2011

Discovery of fourth moon of Pluto: Kerberos.

📷 2011

Photograph of Pluto by Hubble Space Telescope.

2012

Discovery of fifth moon of Pluto: Styx.

📷 2012

Photograph of Pluto by Hubble Space Telescope.

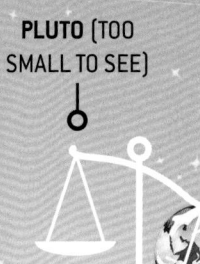

PLUTO (TOO SMALL TO SEE)

MOONS: 1
CHARON

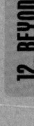

1976

A new estimate of Pluto's **ALBEDO** (the amount of sunlight it reflects) concludes it is bright but small. Mass revised to 0.01x Earth.

 ### 1978

Photographs of Pluto show a bulge that disappears — the first sighting of its moon, Charon. The discovery enables a more accurate calculation of mass: 0.0015x Earth.

1992

Discovery of 15760 Albion, initially called a "minor planet" and the first of more than 2,400 icy objects since found beyond Neptune.

MOONS: 3

CHARON

HYDRA

NIX

OCTOBER 2005

Two more moons of Pluto discovered: Nix and Hydra.

JULY 2005

Discovery of Eris, a planet-like object larger than Pluto.

 ### 1996

Photograph of Pluto by Hubble Space Telescope.

 ### 1994

Photograph of Pluto by Hubble Space Telescope.

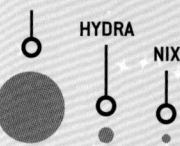

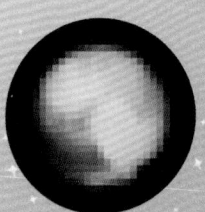

 **APRIL 2015**

Photograph of Pluto by New Horizons.

 MAY 2015

Photograph of Pluto by New Horizons.

 JUNE 2015

Photographs of Pluto by New Horizons.

JULY 2015

New Horizons arrives at Pluto, providing our first detailed look at the mysterious world.

DWARF PLANETS

The five not-quite planets of the Solar System.

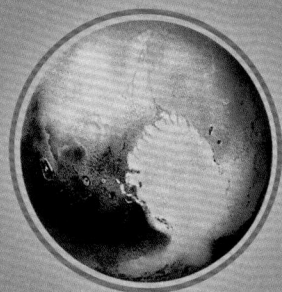

PLUTO

* Discovered 1930, considered to be a planet for 76 years
* 5 known moons

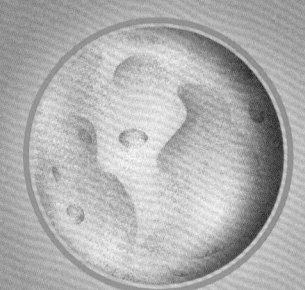

ERIS

* Discovered 2005, prompted IAU to agree on definitions of "planet" and "dwarf planet"
* Smaller but more massive than Pluto
* 1 known moon

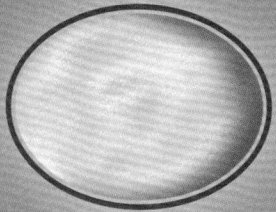

HAUMEA

* Discovered 2004
* 2 known moons

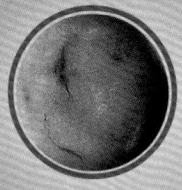

MAKEMAKE

* Discovered 2005
* Second brightest Kuiper Belt object after Pluto
* 1 known moon

GONGGONG

* Discovered 2007
* In orbital resonance with Neptune
* 1 known moon

QUAOAR

* Discovered 2002
* 1 known moon

SEDNA

* Discovered 2003

INCLINATION

How tilted are the Dwarf Planet's orbits?

MAKEMAKE
(45.4 AU, 29°)

HAUMEA
(43.1 AU, 28.2°)

ORCUS
(39.2 AU, 20.6°)

PLUTO
(39.5 AU, 17.1°)

SEDNA
(506 AU, 11.9°)

Approximately 12 pages aw

CERES
(2.8 AU, 10.6°)

SUN

SATURN
(9.6 AU, 2.5°)

URANUS
(19.2 AU, 0.8°)

NEPTUNE
(30.1 AU, 1.8°)

QUAOAR
(43.7 AU, 8°)

ECLIPTIC PLANE

ERIS
(67.9 AU, 44.0°)
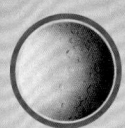

DWARF PLANET CANDIDATES

Other candidates for dwarf planet status include **SALACIA, VARUNA, IXION, 2003 AZ84, 2004 GV9** and **2002 AW**.

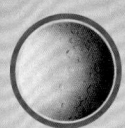

CERES

* Discovered 1801, considered to be a planet for about 60 years, then considered to be an asteroid for about 145 years

GONGGONG
(67.5 AU, 30.6°)

ORCUS

* Discovered 2004
* 1 known moon

WHAT IS A DWARF PLANET?

According to the International Astronomical Union resolution agreed in August 2006, a dwarf planet is a celestial body that:

1. Orbits the Sun
2. Is massive enough to be (nearly) spherical
3. Has NOT cleared the neighbourhood round its orbit (as a PLANET does)
4. Is not a satellite or moon of another body

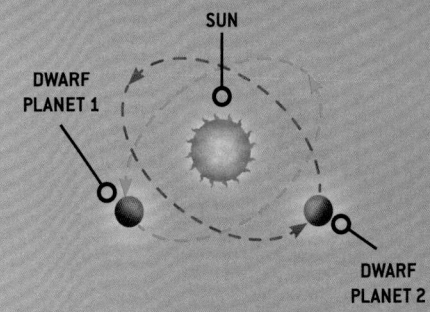

ORBITAL RESONANCE

Two dwarf planets and two dwarf planet candidates are in orbital resonance with Neptune: their gravitational effect on each other means that in the same time…

SEDNA →

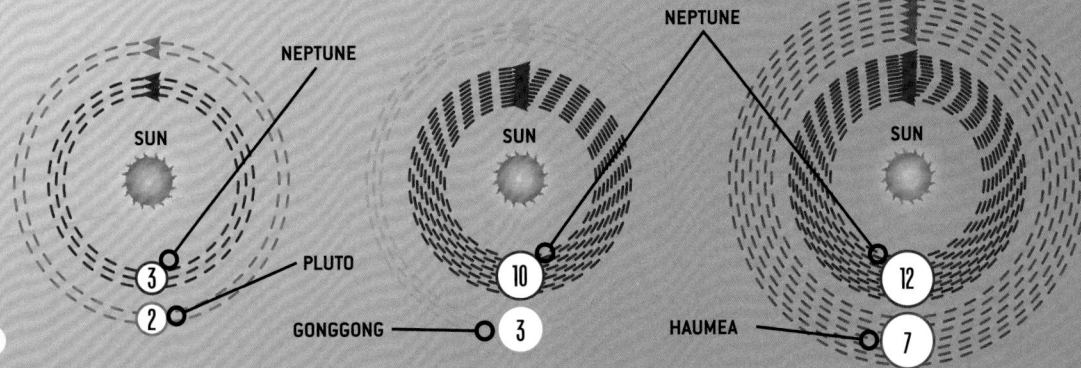

…**Neptune** completes 3 orbits, **Pluto** completes 2 orbits (**Neptune** is also in 3:2 orbital resonance with **Ixion**)

Neptune completes 10 orbits, **Gonggong** completes 3 orbits.

Neptune completes 12 orbits, **Haumea** completes 7 orbits.

KUIPER BELT AND OORT CLOUD

The ice and rocks beyond Neptune.

AU FOR PLANETS AND DWARF PLANETS GIVEN AT APHELION, WHEN THEY ARE EACH AT THE POINT IN THEIR ORBITS FURTHEST FROM THE SUN

KUIPER BELT

This image shows the Solar System including the Kuiper Belt. All distances are shown to scale, while the sizes of the planets are not.

S C A T T E R E D   D I S K

SCATTERED DISK

⟵ ⟶

Extends from the edge of the main Kuiper Belt to 1,000 AU (nearly nine more pages)!

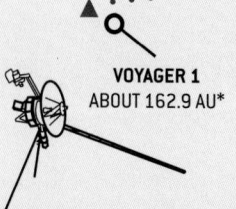

VOYAGER 1
ABOUT 162.9 AU*

HELIOPAUSE

Heliopause, where solar wind no longer strong enough to hold back interstellar radiation.

SCATTERED DISK OBJECTS

Interaction with Neptune's gravity has given these objects highly inclined and eccentric orbits. They include the dwarf planet **Eris**.

OORT CLOUD

We have not yet detected any objects in the Oort Cloud — they are too small and far away — but believe it exists as it seems to be the source of many comets.

DETACHED OBJECTS

Bodies between the scattered disk and the inner Oort Cloud.

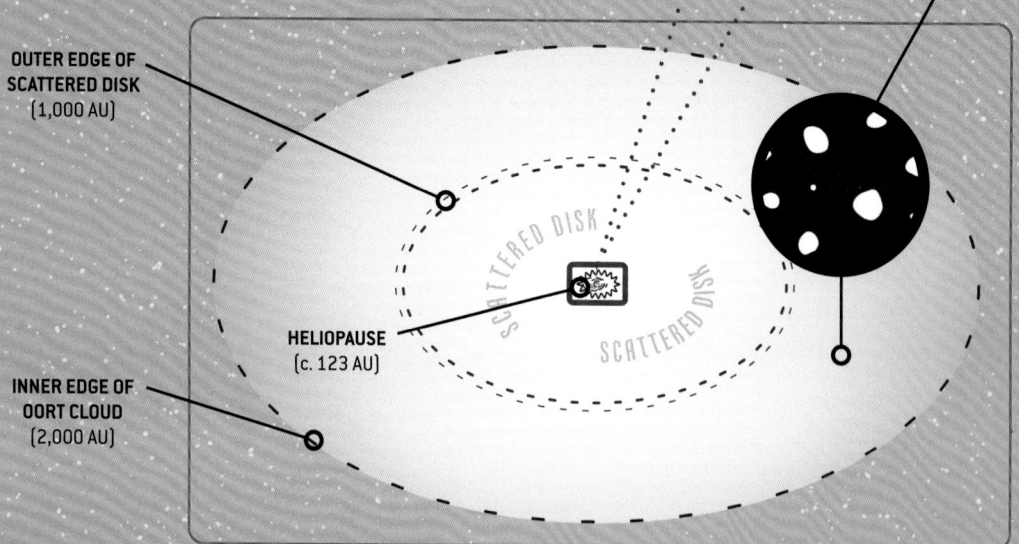

OUTER EDGE OF
SCATTERED DISK
(1,000 AU)

SCATTERED DISK

HELIOPAUSE
(c. 123 AU)

SCATTERED DISK

SCATTERED DISK

INNER EDGE OF
OORT CLOUD
(2,000 AU)

This image shows the Solar System including the Oort Cloud. All distances are shown to scale, while the sizes of the planets are not.

CLASSICAL KUIPER BELT OBJECTS (KBOs)

Ice-rock objects in low-inclination orbits, including the dwarf planets **Pluto** and **Makemake** and smaller objects such as **Varuna** and **Quaoar**.

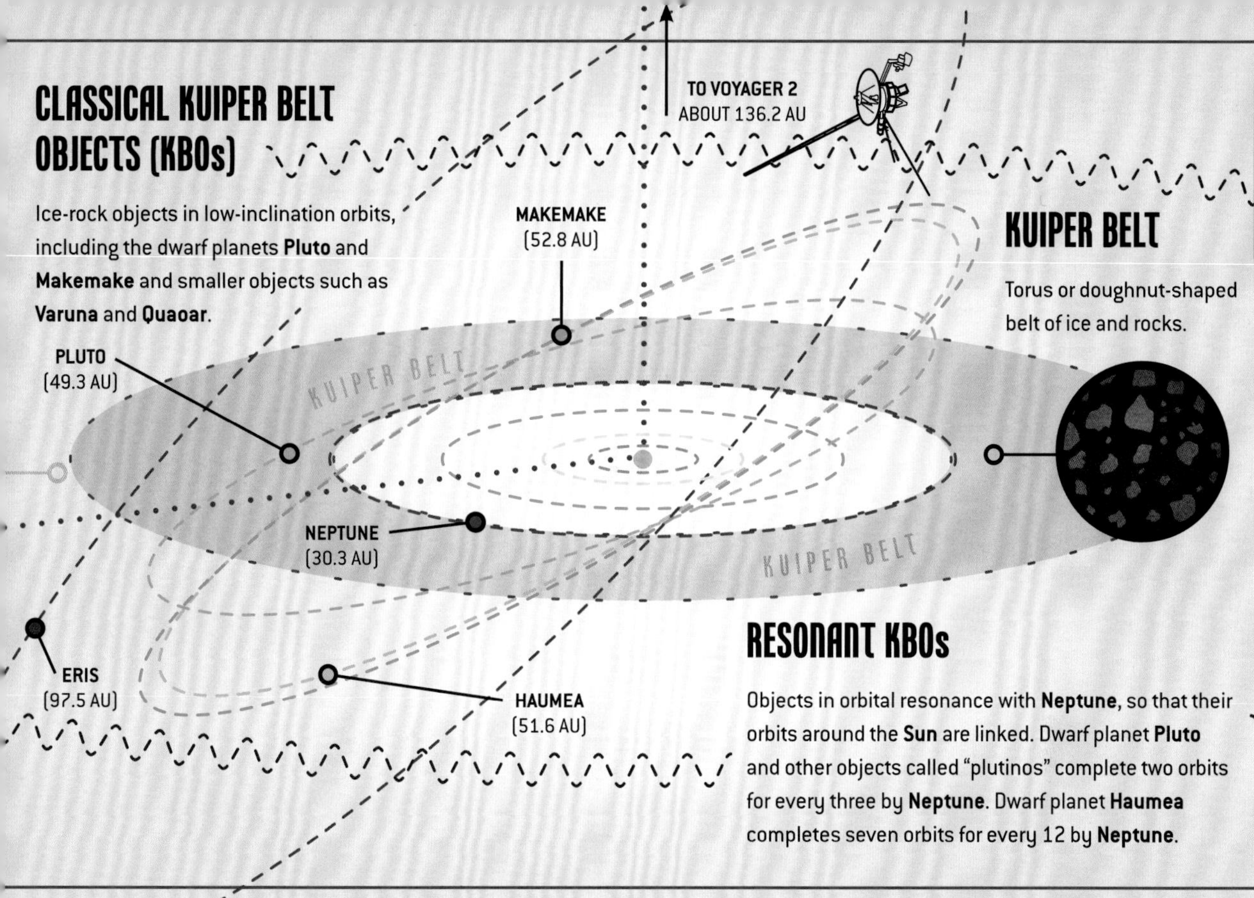

TO VOYAGER 2
ABOUT 136.2 AU

MAKEMAKE
(52.8 AU)

PLUTO
(49.3 AU)

KUIPER BELT

NEPTUNE
(30.3 AU)

ERIS
(97.5 AU)

HAUMEA
(51.6 AU)

KUIPER BELT

KUIPER BELT

Torus or doughnut-shaped belt of ice and rocks.

RESONANT KBOs

Objects in orbital resonance with **Neptune**, so that their orbits around the **Sun** are linked. Dwarf planet **Pluto** and other objects called "plutinos" complete two orbits for every three by **Neptune**. Dwarf planet **Haumea** completes seven orbits for every 12 by **Neptune**.

OORT CLOUD
(Billions of comets surrounding our solar system)

COMET

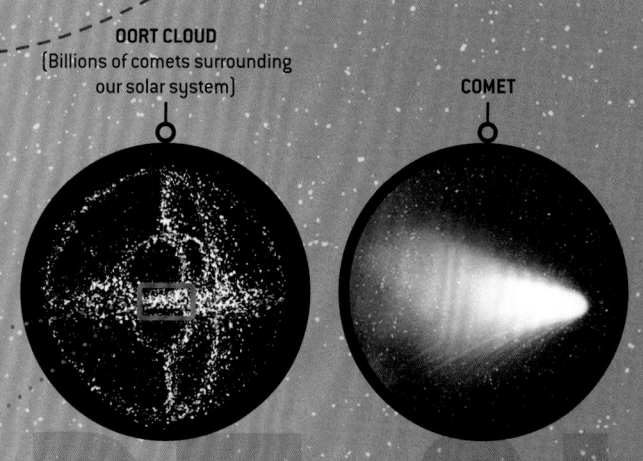

OUTER EDGE →

At 100,000 AU, the outer edge of the Oort Cloud is at the limit of the Sun's gravitational influence. If shown, the edge would be another 27 pages away!

SHELL OF OBJECTS

Home to about 1,000,000,000,000 irregular bodies of ice and rock, dust and organic molecules.

Some comets can be affected by the gravity of gas planets such as Jupiter. Their orbits become smaller and closer, within the Kuiper Belt and lasting <200 years. These are **short-period comets**.

Gravitational interactions cause some of these objects to "fall" towards the Sun and back, in orbits of >200 years. These are **long-period comets**.

See JUPITER > COMETS

VOYAGER 2

The probe, Voyager 2, won't reach the Oort Cloud for 300 years! It will then take c. 30,000 years to reach the far side!

ANOTHER SUN, ANOTHER EARTH

Beyond the Solar System are yet more stars and planets...

Distances from our Sun to nearest stars and known planets:

PROXIMA CENTAURI – 4.2 LIGHT YEARS

* Planet 0.9x mass of Earth but orbits 0.05 AU from its star so unlikely to be habitable

BARNARD'S STAR – 5.6 LIGHT YEARS

* Planet 4.2x mass of Earth

LUHMAN 16 – 6.5 LIGHT YEARS

WISE 0855-0714 – 7.3 LIGHT YEARS

WOLF 359 – 7.9 LIGHT YEARS

* Planet 3.8x mass of Earth
* Planet 43.9x mass of Earth

LALANDE 21185 – 8.3 LIGHT YEARS

* Planet 2.9x mass of Earth

SIRIUS – 8.7 LIGHT YEARS (the brightest star in Earth's sky after the Sun)

LUTYEN 726-8 – 8.8 LIGHT YEARS

ROSS 154 – 9.7 LIGHT YEARS

ROSS 248 – 10.3 LIGHT YEARS

EPSILON ERIDANI – 10.4 LIGHT YEARS

* Planet 490x mass of Earth

LACAILLE 9352 – 10.7 LIGHT YEARS

* Planet 4.1x mass of Earth
* Planet 9x mass of Earth
* Planet 6.5x mass of Earth

ROSS 128 – 11 LIGHT YEARS

* Planet 1.4x mass of Earth and in habitable zone: nearest Earth-like planet

OORT CLOUD, OUTER EDGE

OORT CLOUD, INNER EDGE

SUN

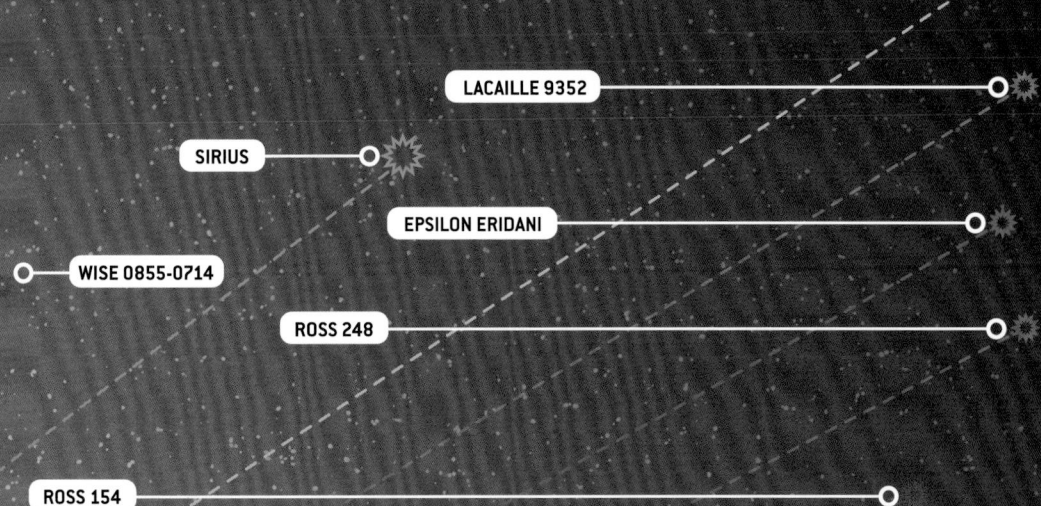

ROSS 128

LACAILLE 9352

SIRIUS

EPSILON ERIDANI

WISE 0855-0714

ROSS 248

ROSS 154

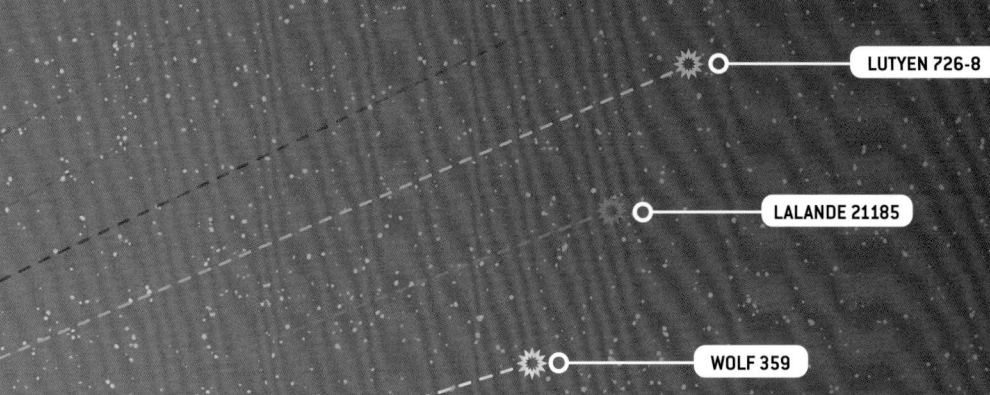

LUTYEN 726-8

LALANDE 21185

WOLF 359

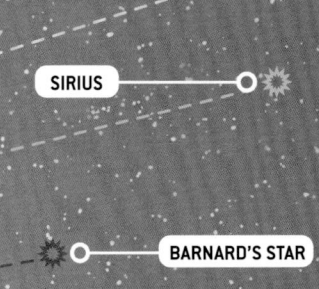

SIRIUS

BARNARD'S STAR

PROXIMA CENTAURI

LIGHT YEARS

At this scale we use a new measure of distance: the light year, or the distance light can travel in one Earth year of **365.25 days**.

SPEED OF LIGHT (c) = 299,792,458 m/s

1 LIGHT YEAR =

* 9,460,000,000,000 km
* 63,241 AU
* 0.3 parsecs (used to measure even greater scales!)

Distance from our Sun to outer edge of Oort Cloud: **100,000 AU** or **1.6 light years**

EXOPLANETS

*The 5,577 planets we've found orbiting **other** stars...*

The first confirmed detection of an "exoplanet" – orbiting a star other than the Sun – was in 1992.

Up to the beginning of 2024, 5,577 exoplanets have been confirmed, with 4,114 planetary systems — and 887 stars other than the Sun known to have more than one planet.

Between 2009 and 2018, the space telescope Kepler discovered 2,662 exoplanets via transit photometry!

HOW WE FIND EXOPLANETS

234 found by IMAGING with telescopes

This isn't easy. Stars are 1 billion times brighter than planets orbiting round them in visible light, and 1 million times brighter when viewed in infra-red.

23 found by ASTROMETRY – wobbles in the positions of stars

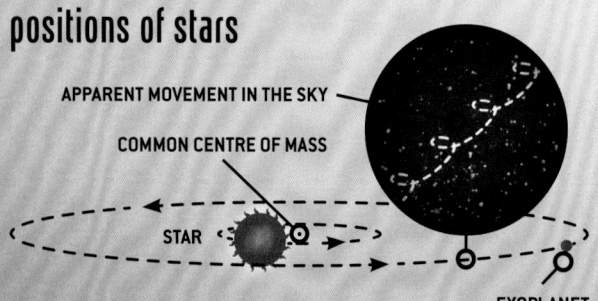

APPARENT MOVEMENT IN THE SKY

COMMON CENTRE OF MASS

STAR

EXOPLANET

A planet orbiting a star causes that star to move in a tiny ellipse.

That small, regular wobble can be measured with respect to other, non-wobbling stars. However, such wobbles are extremely small.

1,084 found by RADIAL VELOCITY – wobbles in the colour of starlight

If a light source such as a star moves towards us, the light appears more blue. If the star moves away, its light appears more red. This is called **DOPPLER SHIFT**.

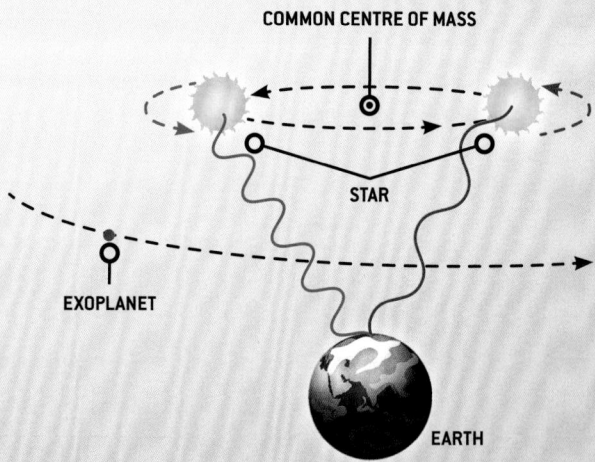

COMMON CENTRE OF MASS

STAR

EXOPLANET

EARTH

As with astrometry, we can look for regular wobbles in Doppler shift. This is easier than astrometry — and more sensitive, able to detect changes as low as 1 m/s.

As a result, radial velocity can give the period or orbit (the length of the planet's year) and a lower limit on the planet's mass. It can also detect several planets orbiting the same star. But it has limitations and is best at detecting massive planets that orbit close to their star (so-called "Hot Jupiters").

3,863 found by TRANSIT PHOTOMETRY – dips in starlight

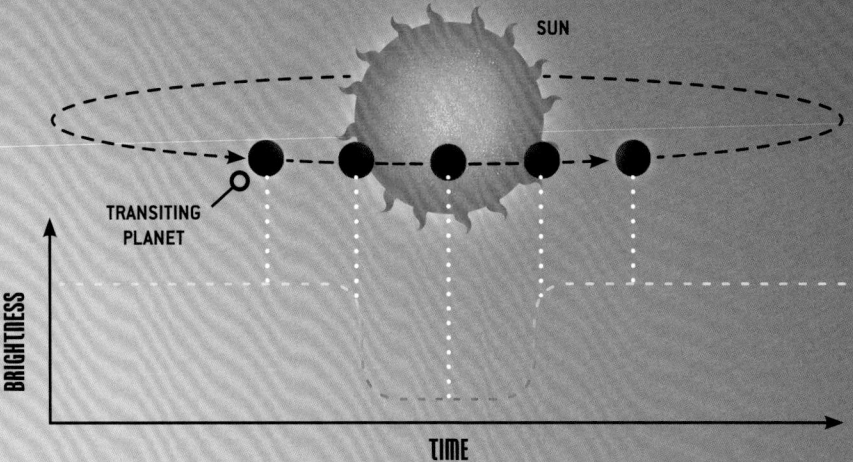

SUN

TRANSITING
PLANET

BRIGHTNESS

TIME

A Jupiter-sized planet transiting (passing in front of) its star blocks 0.01% of that star's light. If the orbital plane of the planet is edge-on to us, we can measure these dips in light and deduce the period of orbit, size of planet and the chemical elements of any planetary atmosphere.

Finding H_2O, O_2, O_3, CO_2, CO or CH_4 might suggest life was possible there. Detecting changes, such as variations in CH_4 and O_2, would suggest biological processes — life!

3,863 exoplanets have been found this way but it also produces lots of false-positive results. These can then be confirmed (or not) by radial velocity.

We can combine the two methods. Knowing both an exoplanet's mass (from radial velocity) and size (from transit method), we can deduce its bulk density and estimate whether it's a rocky world, gas giant or ice giant etc.

267 Found by MICROLENSING

The gravity of the star and its exoplanet bend light coming from another, more distant star as it passes behind them.

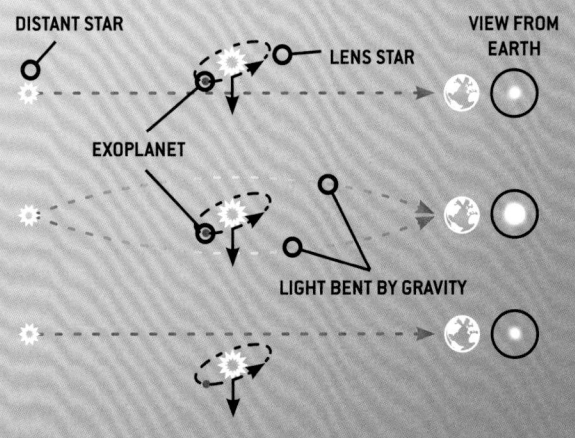

DISTANT STAR

LENS STAR

VIEW FROM EARTH

EXOPLANET

LIGHT BENT BY GRAVITY

ORBITAL BRIGHTNESS MODULATIONS

Massive planets in close orbit can affect the shape of their star.

PULSAR TIMING VARIATIONS

Small, dense neutron stars called "pulsars" produce extremely regular radio pulses, so any variation suggests a planet in orbit. This method had found the smallest known exoplanets — 1/10th the mass of Earth.

29 found by TRANSIT TIMING VARIATIONS

Variations in the time a discovered exoplanet takes to transit its star can reveal other exoplanets in the same system.

PULSATION TIMING VARIATIONS

Some stars regularly pulsate, and a planet can affect those pulsations.

ECLIPSE TIMING VARIATIONS

Binary stars orbiting around one another will eclipse each other at varying times if there is also a planet in orbit.

DISK KINEMATICS

Studying the structure of the disk of gas around a young star suggests it has a newly formed planet.

FURTHER AND FURTHER OUT

*The Solar System might seem enormous but the space
beyond is vastly, hugely, mind-bogglingly big …*

1. SOLAR SYSTEM

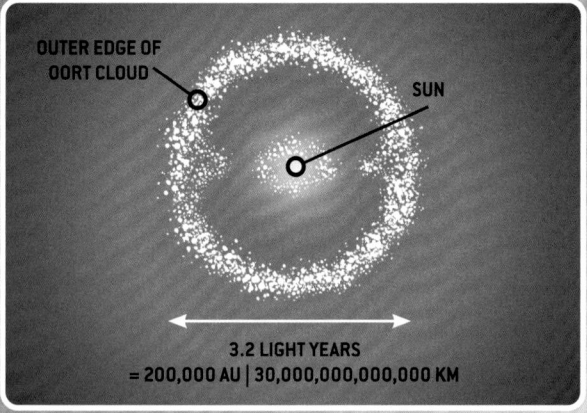

OUTER EDGE OF
OORT CLOUD

SUN

3.2 LIGHT YEARS
= 200,000 AU | 30,000,000,000,000 KM

2. LOCAL INTERSTELLAR CLOUD (LIC)

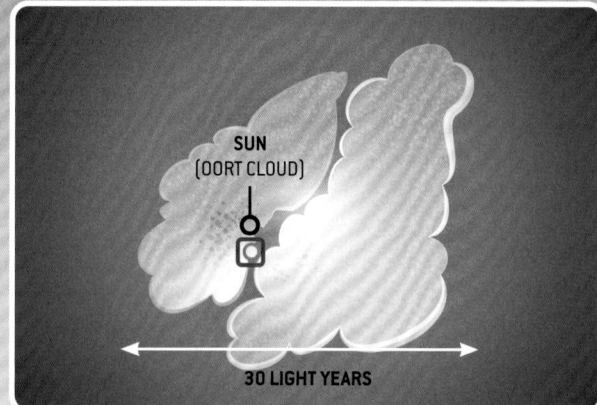

SUN
(OORT CLOUD)

30 LIGHT YEARS

3. LOCAL BUBBLE

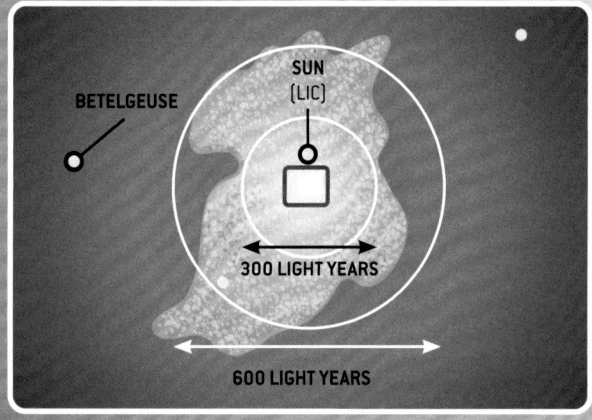

BETELGEUSE

SUN
(LIC)

300 LIGHT YEARS

600 LIGHT YEARS

4. GOULD BELT

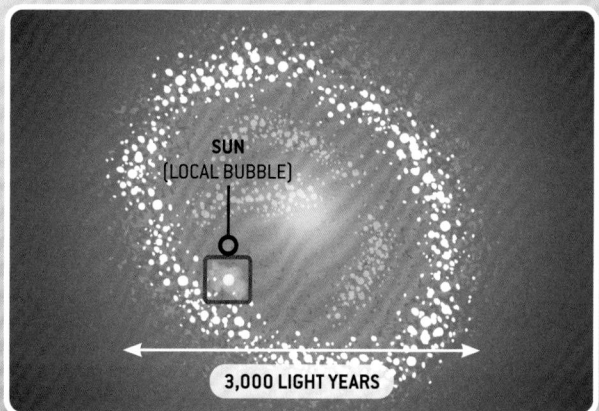

SUN
(LOCAL BUBBLE)

3,000 LIGHT YEARS

5. ORION ARM

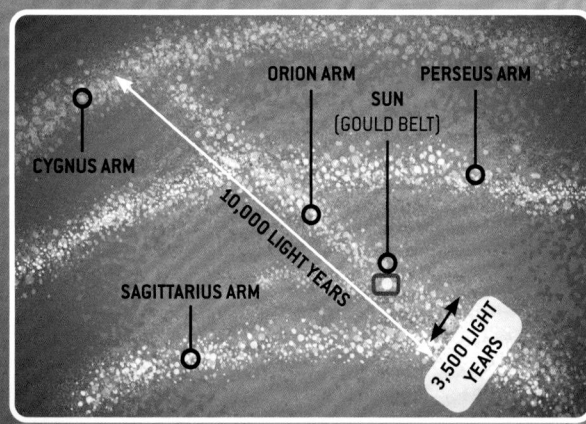

CYGNUS ARM

ORION ARM

PERSEUS ARM

SUN
(GOULD BELT)

10,000 LIGHT YEARS

SAGITTARIUS ARM

3,500 LIGHT YEARS

6. MILKY WAY GALAXY

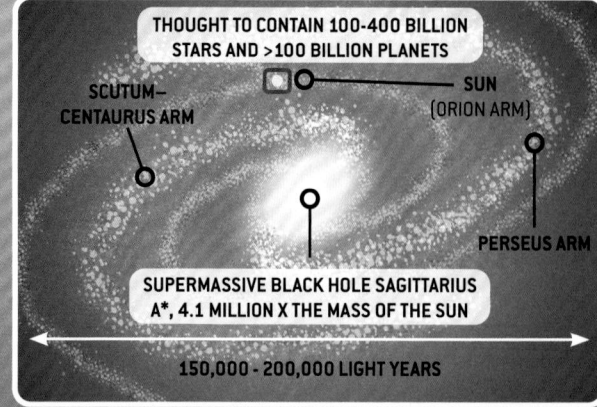

THOUGHT TO CONTAIN 100-400 BILLION
STARS AND >100 BILLION PLANETS

SCUTUM–
CENTAURUS ARM

SUN
(ORION ARM)

PERSEUS ARM

SUPERMASSIVE BLACK HOLE SAGITTARIUS
A*, 4.1 MILLION X THE MASS OF THE SUN

150,000 - 200,000 LIGHT YEARS

7. LOCAL GROUP OF GALAXIES

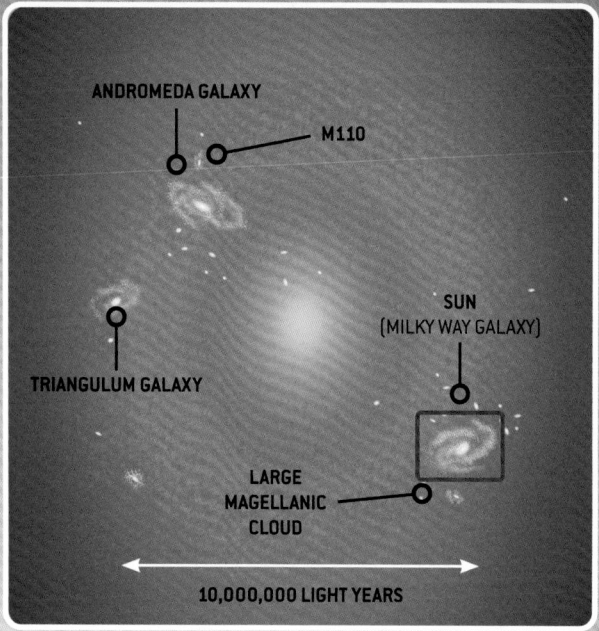

ANDROMEDA GALAXY

M110

TRIANGULUM GALAXY

SUN
(MILKY WAY GALAXY)

LARGE
MAGELLANIC
CLOUD

10,000,000 LIGHT YEARS

8. VIRGO SUPERCLUSTER

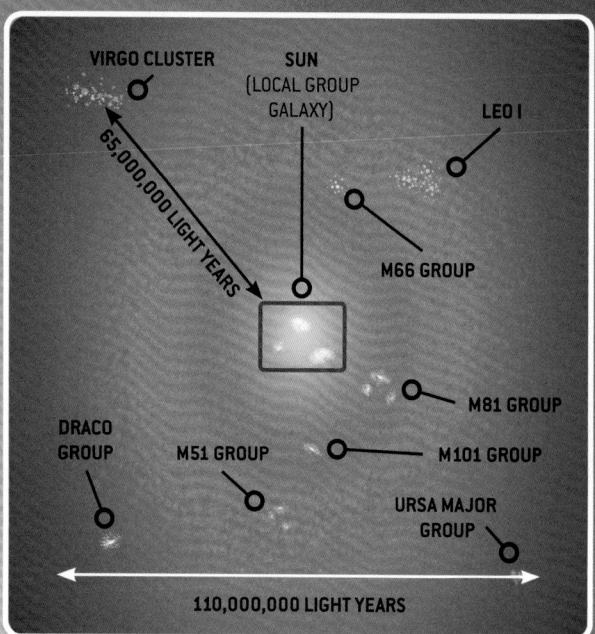

VIRGO CLUSTER

SUN
(LOCAL GROUP
GALAXY)

LEO I

65,000,000 LIGHT YEARS

M66 GROUP

M81 GROUP

DRACO
GROUP

M51 GROUP

M101 GROUP

URSA MAJOR
GROUP

110,000,000 LIGHT YEARS

9. LANIAKEA SUPERCLUSTER

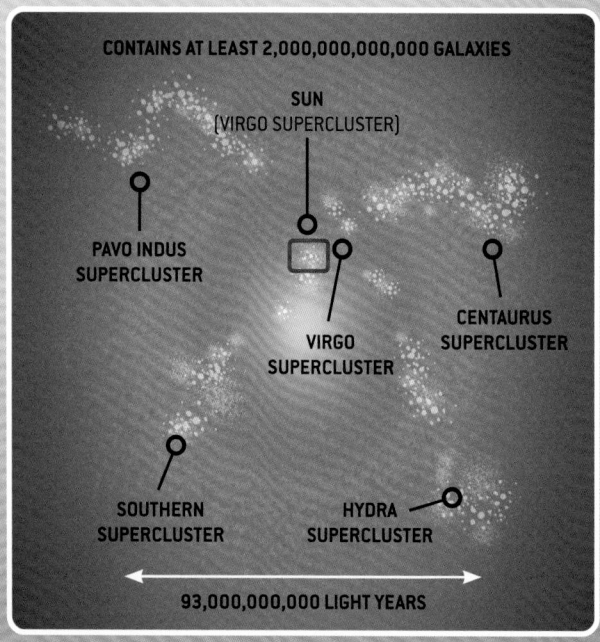

CONTAINS AT LEAST 2,000,000,000,000 GALAXIES

SUN
(VIRGO SUPERCLUSTER)

PAVO INDUS
SUPERCLUSTER

CENTAURUS
SUPERCLUSTER

VIRGO
SUPERCLUSTER

SOUTHERN
SUPERCLUSTER

HYDRA
SUPERCLUSTER

93,000,000,000 LIGHT YEARS

10. OBSERVABLE UNIVERSE

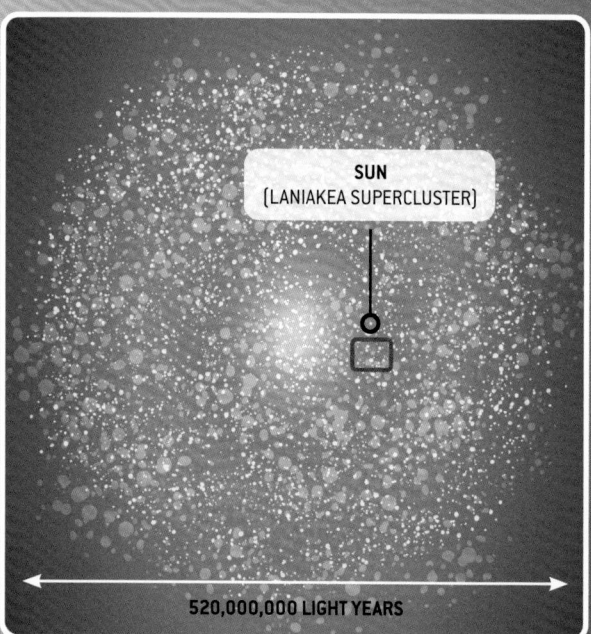

SUN
(LANIAKEA SUPERCLUSTER)

520,000,000 LIGHT YEARS

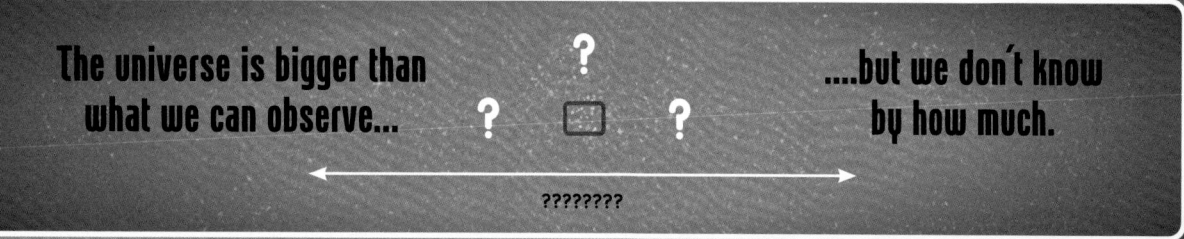

The universe is bigger than
what we can observe...

?

?   ?

....but we don't know
by how much.

????????

GLOSSARY

Understanding the solar system can involve the use of lots of terminology. In this section we translate this into something more accessible...

ABERRATION (OPTICS)

Optical aberration refers to imperfections in the formation of an image by optical systems such as lenses or mirrors. These imperfections cause the image produced by the system to deviate from the ideal or mathematically calculated form.

ABERRATION (ASTRONOMY)

Astronomical or stellar aberration is a phenomenon that occurs due to the combination of the finite speed of light and the motion of an observer relative to objects being observed. The apparent positions of bodies appear slightly shifted from their true positions.

See **SATURN > VISIONS OF SATURN**  133

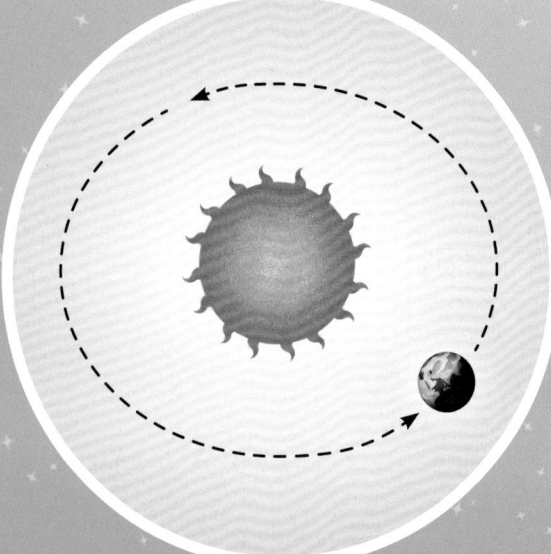

CLEARING THE NEIGHBOURHOOD

This means that a celestial body has become gravitationally dominant within its orbital zone by either capturing or scattering objects that might otherwise share its orbital path.

See **MERCURY > WHAT IS A PLANET?**
 27

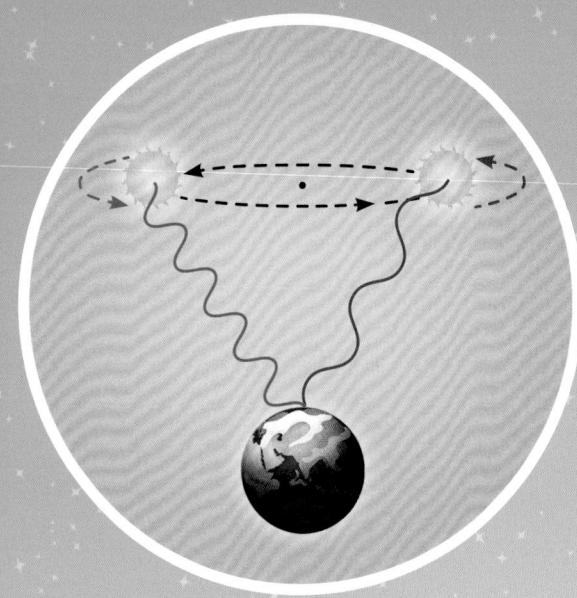

DOPPLER SHIFT

This refers to the change in the observed wavelength of light emitted by a object due to the relative motion between the object and the observer.

The wavelengths emitted by that object can appear shifted, either towards the blue if the object is moving towards the observer, or towards the red if the object is moving away from the observer.

See **BEYOND > EXOPLANETS**

172

EJECTA

Material thrown out (or ejected) from a volcano or as an impact crater is formed.

See **MERCURY > TITLE PAGE**

22

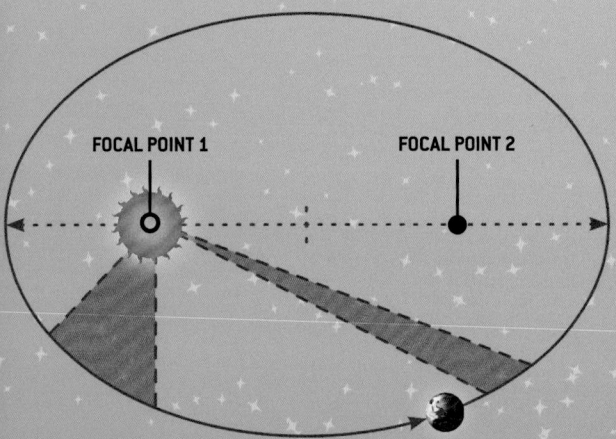

FOCAL POINT 1   FOCAL POINT 2

FOCAL POINT

The two centres of an orbit. In a perfectly circular orbit, the two focal points are in the same place: the middle of the circle. In elliptical orbits, the focal points are separated: and the more separated, the less circular the orbit. In a planetary orbit, the Sun is at one of the focal points.

See **URANUS > PLANETARY MOTION**

150

GLANCING

Colliding objects don't always hit each other head on. A low-angle or "glancing" side impact does not deliver the full effect of the blow.

See **MOON > ORIGINS**

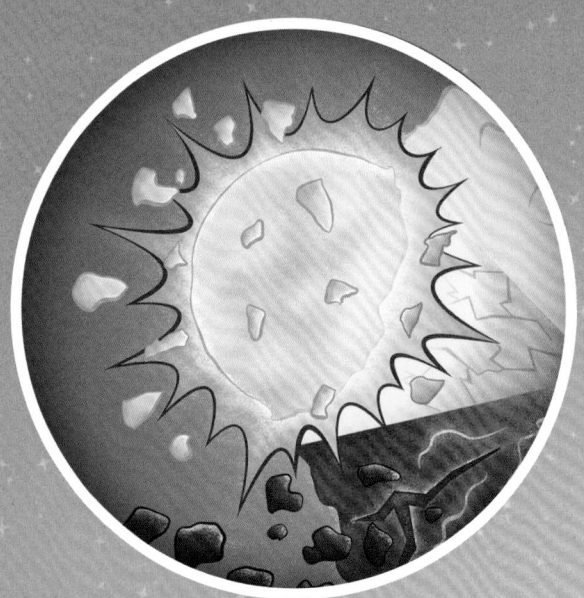

GRAVITY

The attractive force between objects. All objects with mass produce a gravitational field and the more massive the object, the greater its gravitational field. The strength of a gravitational field is measured in newtons per kilogram (N/kg). On Earth's surface, the field strength is 9.8 N/kg.

See **EARTH > TIDES**

LATITUDE

On planets including Earth, a system for indicating a position relative to the equator (0° latitude) and the poles — North (90°N) and South (90°S). Astronomers also use "celestial latitude" to position objects such as stars relative to the ecliptic*.

See **SUN > SUN SPOTS**

*For more about the ecliptic, see **SUN > THE WAY YOU MOVE**

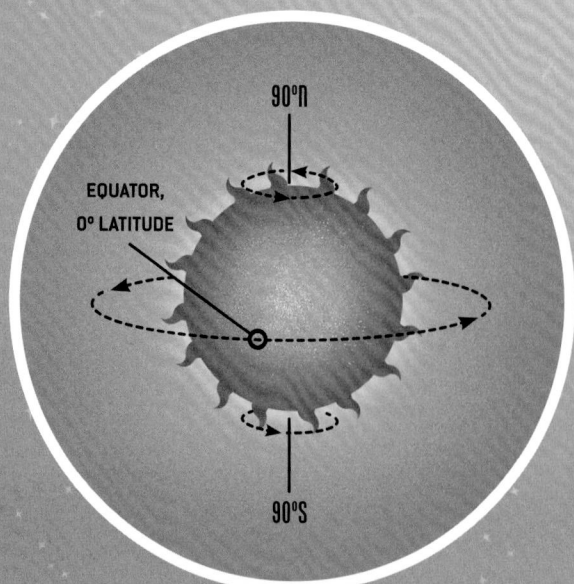

90°N

EQUATOR, 0° LATITUDE

90°S

LUMINOSITY

How bright an object such as a star really is, rather than how bright it appears to us — because it can get dimmer with distance.

See **SUN > JUST MY TYPE**

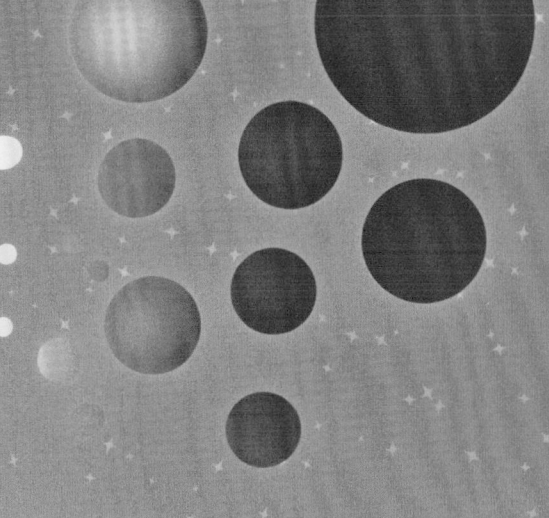

OBLIQUITY

Tilt, the angle by which the rotational axis of a planet or other object is inclined.

See **MANY PAGES!**

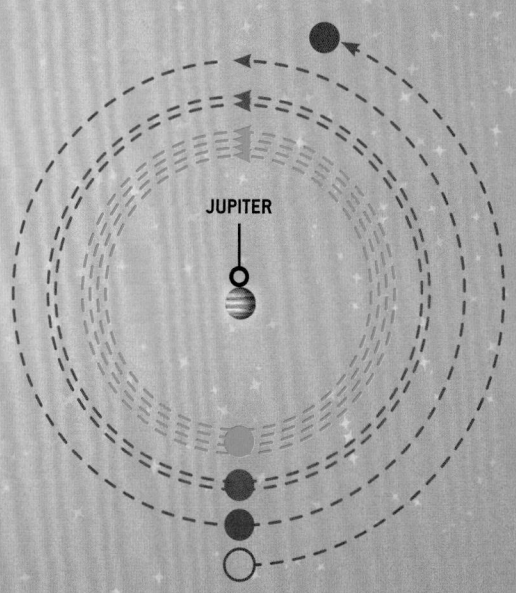

JUPITER

ORBITAL RESONANCE

In certain cases, gravitational forces can "lock" different bodies orbiting the same larger body into step. For example, a planet might have two moons in orbital resonance, one orbiting the planet twice for every four orbits of the other.

See **JUPITER > GALILEAN MOONS, BEYOND > DWARF PLANETS & KUIPER BELTS AND OORT CLOUD**

PERMAFROST

Permanently frozen surface of a planet or other body.

See **MARS > TITLE PAGE**

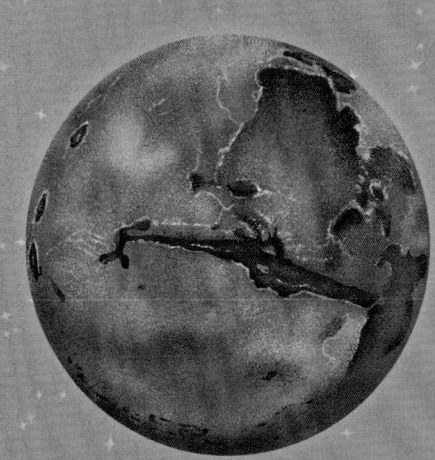

PERTURBED

When an object (e.g. a planet) is deflected from what should be a regular orbit round a larger body (e.g. a star) by the gravitational force of another object (e.g. a second, nearby planet).

See **NEPTUNE > GRAVITY AND THE DISCOVERY OF NEPTUNE**

 154

RADIAL VELOCITY

The component of an object's velocity that is in the direction of the line of sight of an observer. In astronomy, it specifically refers to the motion of a celestial object either towards or away from the observer. These movements give rise to doppler shifts towards the red (if the object is moving away) or blue (if the object is moving towards) the observer.

See **BEYOND > EXOPLANETS & GLOSSARY > DOPPLER SHIFT**

 172   177

PROMINENCES

A large, outward eruption of gas from the Sun's surface, often in a looped shape and especially bright when viewed (from Earth) during a solar eclipse.

See **SUN > SUNSPOTS & MOON > ECLIPSES**

 18   68

RAREFIED

Less dense or thinner than usual, especially when referring to gas or atmosphere.

See **SUN > TITLE PAGE, MERCURY > TITLE PAGE & JUPITER > ATMOSPHERES**

 10   22   120

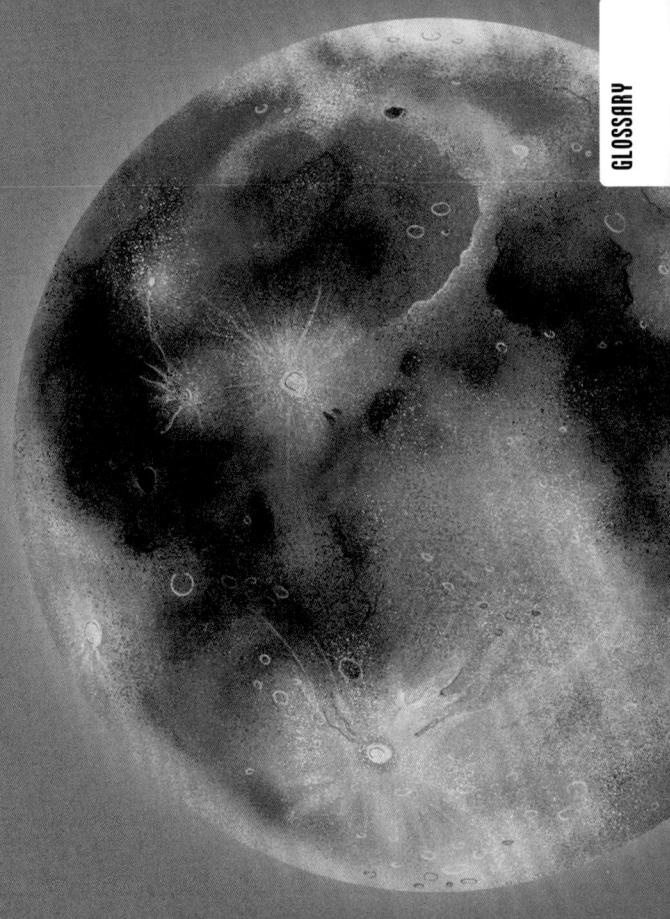

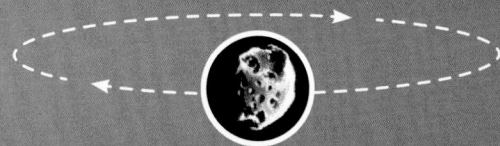

RETROGRADE

In the opposite direction to usual. For example: most stars and objects move west to east (prograde) across the sky but retrograde objects move east to west.

See **SATURN > MOON TYPES – II & URANUS > URANIAN MOONS**

143   149

ROTATION PERIOD

The time taken for a body to rotate once on its axis; a day.

See **SUN > TITLE PAGE & MOON > TITLE PAGE**

10   62

SILICATE ROCKS

A kind of mineral that makes up 90% of Earth's crust, composed of silicon and oxygen.

See **MOON > TITLE PAGE & JUPITER > GALILEAN MOONS**

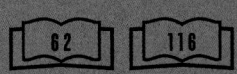

62   116

VOLCANISM

The action of a volcano.

See **VENUS > TITLE PAGE & MOON > TITLE PAGE**

32   62

INDEX

ACKNOWLEDGEMENTS

AS A DYSLEXIC AUTHOR, WORKING AS A TEAM IS INVALUABLE, SO I WOULD LIKE TO GIVE A BIG SHOUT OUT TO THE MEMBERS WHO HELPED BRING THIS BOOK TO LIFE.

Firstly, to **SIMON**, my co-writer in this adventure, who would in turn like to thank Dr Debbie Challis, Ben Morris, Steve O'Brien, the Lord of Chaos and Lady Vader.

EMMA, the brilliant designer and illustrator behind this book, would like to thank Alex, Alan and Marianne, Sherrie and Ian, and her own little rocket-lovers Oscar and Sylvie. Most importantly she would like to thank her husband, best friend and space explorer, Si.

For advice, assistance and inspiration and the production of the glorious images that this book contains, we would like to thank:

Simon Belcher, Dallas Campbell, Dr Jessie Christiansen, Dr David Hone, Tom Kerss, Timothy Knapman, David McCandless, Randall Munroe, Brendan Owen, Professor Michael Scott, Dr Radmila Topalovic and Dr Zoe Williams.

As an author, I am still not quite house trained, so I would like to thank the team at BBC Books for their insight, support and patience throughout this process: Albert, Céline, Charlotte, Dan, Antony and Morgana.

On a personal note, I would also like to thank my wonderful daughter Lori, who is my daily inspiration in all things.

And finally, to all the amazing scientists, engineers and seekers, who work together to find out more about what is out there.

1

BBC Books, an imprint of Ebury Publishing
1 Embassy Gardens, 8 Viaduct Gardens
London, SW11 7BW

BBC Books is part of the Penguin Random House group of companies
whose addresses can be found at global.penguinrandomhouse.com

Penguin
Random House
UK

First published by BBC Books in 2024

www.penguin.co.uk

Commissioning editor: Albert DePetrillo
Editor: Charlotte Macdonald
Project editor: Simon Guerrier
Designer and illustrator: Emma Price

A CIP catalogue record for this book is available from the British Library

ISBN 9781785949203

Printed and bound in China by C&C Offset Printing Co., Ltd.

Penguin Random House is committed to a sustainable future for
our business, our readers and our planet. This book is made
from Forest Stewardship Council® certified paper.

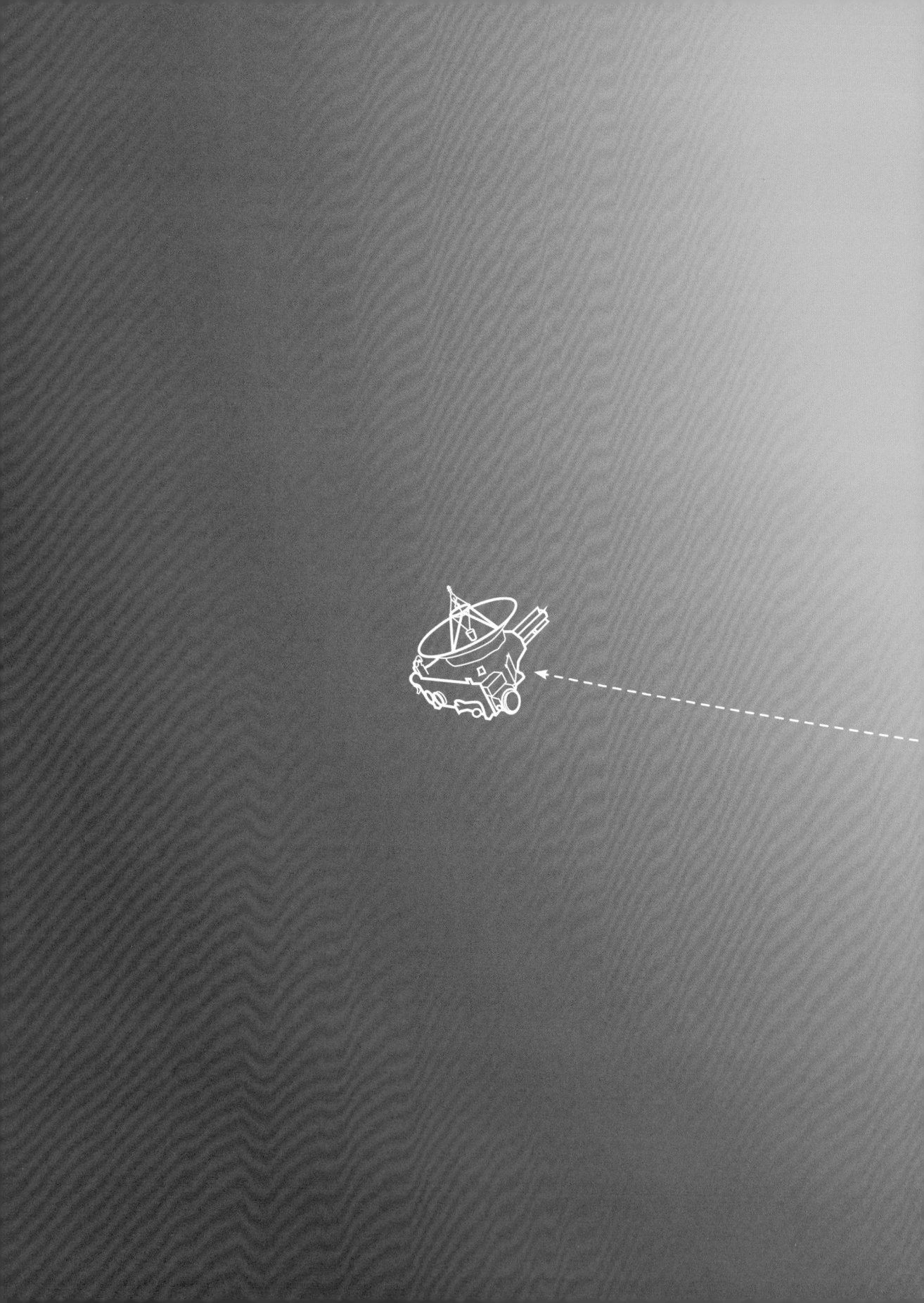